D0722844

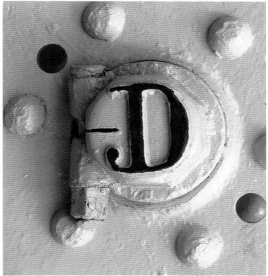

The Farm Tractor

100 Years of North American Tractors

Ralph W. Sanders

Foreword by Bob Feller

Voyageur Press

Dedication

To the several hundred antique and classic tractor owners and restorers across North America that I have had the pleasure to work with over the past eighteen years. Most of their machines are pictured herein.

Their patient submission of their priceless machines to my photography efforts produced these results. I am eternally grateful to them, and my hope is that our combined efforts are worthwhile and meaningful to this generation and to following generations.

— Ralph W. Sanders

First published in 2007 by Voyageur Press, an imprint of MBI Publishing Company, Galtier Plaza, Suite 200, 380 Jackson Street, St. Paul, MN 55101 USA

Voyageur Press titles are also available at discounts in bulk quantity for industrial or sales-promotional use. For details write to Special Sales Manager at MBI Publishing Company, Galtier Plaza, Suite 200, 380 Jackson Street, St. Paul, MN 55101 USA.

To find out more about our books, join us online at www.VoyageurPress.com.

Editor: Leah Noel
Designer: LeAnn Kuhlmann

Printed in Hong Kong
Library of Congress Cataloging-in-Publication Data

Sanders, Ralph W., 1933-
  The farm tractor : 100 years of North American tractors / by Ralph W. Sanders.
    p. cm.
  Includes bibliographical references.
  ISBN-13: 978-0-7603-3074-6 (hardbound w/ jacket)
  ISBN-10: 0-7603-3074-3 (hardbound w/ jacket)
  1. Farm tractors--North America--History. 2. Farm tractors--North America--Pictorial works. I. Title. II. Title: 100 years of North American tractors.
TL233.6.F37S25 2007
629.225'2097--dc22
                    2007020327

On the front cover
The first tractor with power lift for raising attached implements, the 1928 GP (for General Purpose) was designed as a three-row, row-crop machine. An arched front axle and step-down gearing to the rear wheels gave it extra clearance for straddling the center row. The father-and-son team of James and Terry Thompson of Laurelville, Ohio, restored this tractor.

On the frontis page
The tractors that built the American agricultural industry came from a variety of manufacturers, including International Harvester, Oliver Hart-Parr, Caterpillar, John Deere, Rumely, and Wallis.

On the title pages
Built big to plow the wide northern prairies, the 1910 Pioneer 30/60 featured a smooth-running opposed four-cylinder gas engine with a 7x8-inch bore and stroke. Irvin King of Artesian, South Dakota, gathered this nearly one hundred-year-old tractor's remains from the edge of a field where it had "rested" for fifty-five years and put it back into running condition. The Pioneer Tractor Manufacturing Company of Winona, Minnesota, developed and made this model, a larger 45-horsepower tractor, and three smaller models.

On the back cover
Main: A 1952 red-belly 8N Model, equipped with the Ford 119.7-cubic-inch four-cylinder engine. Ford made more than 524,000 Model 8Ns.
Inset: Ford-Ferguson System of Implements advertisement.
Below: The Big Bud 16V-747, the largest farm tractor made in the twentieth century.

Contents

Acknowledgments

It has been an honor and a privilege over the past eighteen years to photograph a particularly fascinating part of America's proud rural heritage—the faithfully and lovingly restored antique and classic farm tractors from the twentieth century. These sturdy icons of farm mechanization represent a treasured past to those who knew them in their heyday and to those who now collect, restore, and proudly share them with others. I share in the respect and admiration for the early machines that freed our grandfathers, our fathers, and our generation, too, from much arduous farm toil. These machines not only helped raise the standard of living for farm families in the United States, but helped make it a better world with more food and fiber available at less cost. To say that I have enjoyed recording these wonderful machines would be an understatement. One perceptive "old iron" fan, who graciously allowed me to photograph his sparkling relic, observed my glee and total absorption in taking the best possible photo of his proud old machine. "Do you actually accept money for doing this?" he kidded me. When I admitted there *was* a modest fee attached to my "work," he observed, "Then you must have the best job in the world, having all of this fun *and* getting paid for it, too." He was absolutely and totally correct. It has been a real treat! I have loved all of it.

I have photographed farm tractors for more than just the past eighteen years, when I really began to follow these historic antiques. My photographic tractor "work" began about thirty-three years ago in 1974 with many freelance photography assignments for Deere & Company's advertising department in Moline, Illinois. That photography work with Deere continued well into the 1980s. I was honored to accompany Deere advertising art directors and ad copy writers to photograph new John Deere farm and industrial equipment at locations across the United States and Canada. The photos we created were used in Deere product brochures and other advertising pieces. There were also similar "shoots" for Massey-Ferguson and several other short-line agricultural equipment companies. So I got acquainted with the "territory" of farm equipment photography.

I owe the dozens of people I worked with at those companies my lasting appreciation and thanks for their experienced guidance, which helped me get started on the right foot in photographing major modern metal. Their good artistic taste and their preferences for showing their machines at their best angle and with the most appealing lighting influenced my own ideas and my own approach to tractor and farm equipment photography. Thanks to dozens of them for sharing their expertise with me.

I have been a tractor "nut" since I was a youngster growing up on a grain farm in Illinois. By age ten, I had learned to operate our family farm's tractors while doing real field work. Those early experiences added to my firsthand knowledge and to my appreciation for the useful productivity of tractors and other farm machines. Yes, I photographed our farm's tractors too, as my interests in photography grew along with those in farming.

Thanks to my parents, John and Zelda Sanders, for giving me life and the opportunity to grow up on a farm where they provided the selfless guidance and training I needed to mature to manhood. Being a good example and having patience were but two of their virtues.

In 1989, I accepted an invitation from John Harvey to photographically illustrate the first nationally distributed DuPont calendar of antique and classic farm tractors. That calendar assignment continued for more than eleven years and resulted in more than 150 tractors being further "preserved" by photography. Thank you, John, for letting me help!

That calendar "work" led to similar calendar photography for Voyageur Press, which has allowed me to continue working the best job in the world well into the twenty-first century.

My wife Joanne travels with me on tractor photo jaunts when her schedule permits and has become expert at weather watching, road map navigating, lugging and loading heavy camera equipment, cleaning tractor tires and wheels, and reflecting light into dark shadows of the shot. I very much appreciate her steady help and companionship, as well as her unflagging patience with me and with the ever-changing weather.

Joanne and I are both very grateful for the wonderful hospitality we have been shown at the hundreds of farms and homes we have visited, which has included most welcome morning cups of coffee and cookies, much-appreciated lunches, and cooling ice tea and lemonade on hot afternoons.

I especially want to thank Michael Dregni, editorial director of Voyageur Press, for the opportunity to share these antique and classic machines and their fascinating histories with you. I first worked with Michael more than ten years ago when he asked me to write and illustrate my first "tractor" book, *Vintage Farm Tractors: The Ultimate Tribute to Classic Tractors*. Through the intervening years, he has patiently guided and helped me through three more books, as well as many antique farm tractor calendars. Thank you immensely, Michael!

The hundreds of collectors whose machines are featured in this book made this book possible and are its real stars. My special thanks to these tractor owners and their willing helpers who patiently washed, polished, moved, started, moved again, hauled, loaded, dusted, polished, painted, and waited with us for the rain to end and the light to get just right for the photos.

I also want to thank the collector-restorers who allowed me to return more than once to their farms, homes, and other places to photograph other great tractors in some fabulous tractor collections. With some of those patient collectors, who own sheds full of prime models of antique tractor specimens, our tractor "shooting" expeditions to their locations became almost an annual event.

The tractor owners also generously shared a wealth of information with me about their machines and referred me to other sources that furthered my education in the powering up of the American farm. I'm very appreciative and also very impressed with the depth of their knowledge and their willingness to share it with me.

I am eternally grateful for the many, many hours they dedicated to taking on and completing their tractor restoration projects, which made the photographs within this book possible. Their investment of time, loving care, years of experience, and money brought these proud farm machines back to "life" from the past. It is the result of their many efforts that I share these photos of many familiar, and even some strange, machines with you.

Thank you all very much!
I hope you enjoy this book.

La Crosse Happy Farmer TRACTOR

IN three short years, the satisfaction of thousands of owners of Happy Farmer Tractors has built the great business and the 16 big factory buildings of the Happy Farmer Tractor organization. Every man who owns a Happy Farmer Tractor boosts for it. His own experience has shown him that the Happy Farmer is the perfect, one man, kerosene burning tractor for the farm of any size.

From start to finish the Happy Farmer Tractor is built for leadership. Every part of it is made in the Happy Farmer plant by the highest grade workmen and the most up-to-date machinery.

So many thousands of farmers want the Happy Farmer that we cannot keep pace with the demand, although we are continually increasing our manufacturing facilities.

The Right Design

That the Happy Farmer with its wide tread and perfect balance, is the right design of tractor, is proved by its use. Experience has shown that this design delivers more power with less weight.

The Happy Farmer turns in its own tracks to right or left with equal ease.

While it is rated at only 12-24 horsepower, it can always be counted upon to deliver much more than this whenever you need it.

Because the Happy Farmer is so simple and because of our great factory, big buying power, and expert organization, we can offer this master tractor for the low price of $1075.

Happy Farmer Tractor Implements

Happy Farmer Tractor Implements give the same satisfaction as the Happy Farmer Tractor. All Happy Farmer Moldboard Plows are automatically controlled by a cord from the driver's seat on the tractor.

The Happy Farmer Disk Harrow is made especially for tractor work. The Happy Farmer Drill is the only proven successful one with automatic patented power-lift and power pressure.

See The Next Demonstration

Performance in the field is a tractor's only test. Watch the Happy Farmer yourself and you will see why it is America's greatest tractor. There is a Happy Farmer distributor in your locality who will be glad to let you know when the next demonstration is to be held. Write us today for his name.

LA CROSSE TRACTOR COMPANY
Department 259   La Crosse, Wisconsin

$1075

Foreword

By Bob Feller, Baseball Hall of Famer and vintage tractor collector

The first tractor on our family farm in Iowa was a J. I. Case crossmotor that my father purchased in the 1920s. Our family's farm was located in the countryside near Van Meter in the south-central part of the state, and that medium-sized Case pulled a three-bottom plow with ease. Together, that tractor worked the land with us for almost a decade.

When it was time to buy a new tractor in the early 1930s, my dad bought the first Caterpillar in Iowa. Not surprisingly, all our farming neighbors told him he was crazy. "It won't work," folks told him. People in our part of the country drove Fordsons or Farmalls, Johnny Poppers or Olivers—tractors with wheels on them. Nobody used a Caterpillar with those crazy crawler treads on them. It simply wasn't *right*.

Well, naturally they were all wrong. That Cat Twenty proved itself on our farm and made a convert of me, and many another farmers.

As I remember, my dad paid $2,300 for that Cat Twenty teamed with a twelve-foot Caterpillar combine. Just two years later, Cat sold that same combine to Deere & Company.

Working our land, I put in many hours at the controls of that Cat Twenty and the combine. They were solid machines that served us well for many years. My fascination with Caterpillars grew from those roots and continues to grow today.

I left the family farm to earn my living throwing baseballs. When I was seventeen years old in 1936, I made my major league debut pitching for the Cleveland Indians against the St. Louis Cardinals. Over the years, I dueled from the pitching mound with some of the all-time greats, batters such as Ted Williams and Joe DiMaggio—just me against them. Some of the veterans of those days said I threw the fastest pitches they had ever seen.

We all took time out from baseball during the World War II years; I served with the U.S. Navy aboard the USS *Alabama* from the summer of 1942 through spring 1945.

I returned to the mound in September 1945, and remained true to the Cleveland Indians until my retirement from baseball in 1956. At the end of eighteen years of throwing fastballs for the Indians, I had a record of 266 wins against 162 losses, a lifetime ERA of 3.25, and 2,581 strikeouts. In 1962, I was elected to the Baseball Hall of Fame on the first ballot.

Despite my achievements on the baseball diamond, a big part of my heart still belongs to the farm fields of my youth. Nostalgia for hallmarks of our roots seems to hit us harder as we grow older. For me, as for many farmers, one of the ties to my youth was the Caterpillar Twenty that I operated as a kid in the 1930s. I decided I wanted to track down another Twenty, which I soon did. Little did I know, but my life as a Cat collector had begun.

Since finding the Twenty, my small Caterpillar collection has continued to grow. It's kind of my own personal Caterpillar "hall of fame" that includes my favorite Cat models: the Twenty, two Tens, a Forty, a Twenty Two, a Twenty Five, a Twenty Eight, and a D4. All in all, I have owned twenty-seven Caterpillars, and I even had the opportunity to drive a D11 on Cat's own proving grounds in East Peoria.

Since then, I've sold off most of my collection, with the exception of my Cat Twenty and Twenty Eight—although I do have my eye out for a D5 and a D6. That fascination for old tractors never leaves you. . . .

You can look at the latest and greatest farm tractor of any brand today and see the history in the machine—the lineage of the steam age, the perfection of the early gas tractors, the development of diesel power, and the creation of implement systems. Driving a tractor today is a lot different than it was when I was a kid back in the 1930s!

I don't think I need to explain the fascination for vintage farm tractors to anyone holding this book in their hands. Author and photographer Ralph Sanders tells the story of the farm tractor's great history and its importance in feeding the world. There have been other good books published on farm tractors in the past and there certainly will be more in the future, but *The Farm Tractor: 100 Years of North American Tractors* stands out in its scope, in-depth coverage, and beautiful photographs and illustrations. I hope tractor fans everywhere enjoy it.

This Caterpillar Model Twenty is like the one former Cleveland Indian baseball speedball pitcher Bob Feller drove as a boy on his father's Iowa farm. Feller, now of Gates Mills, Ohio, has several Caterpillars in his collection. Feller's 1929 Twenty has a curved rear fender, which was common to the model. Feller keeps his Cat collection near his boyhood home in Iowa.

Chapter One

Muscles to Motors

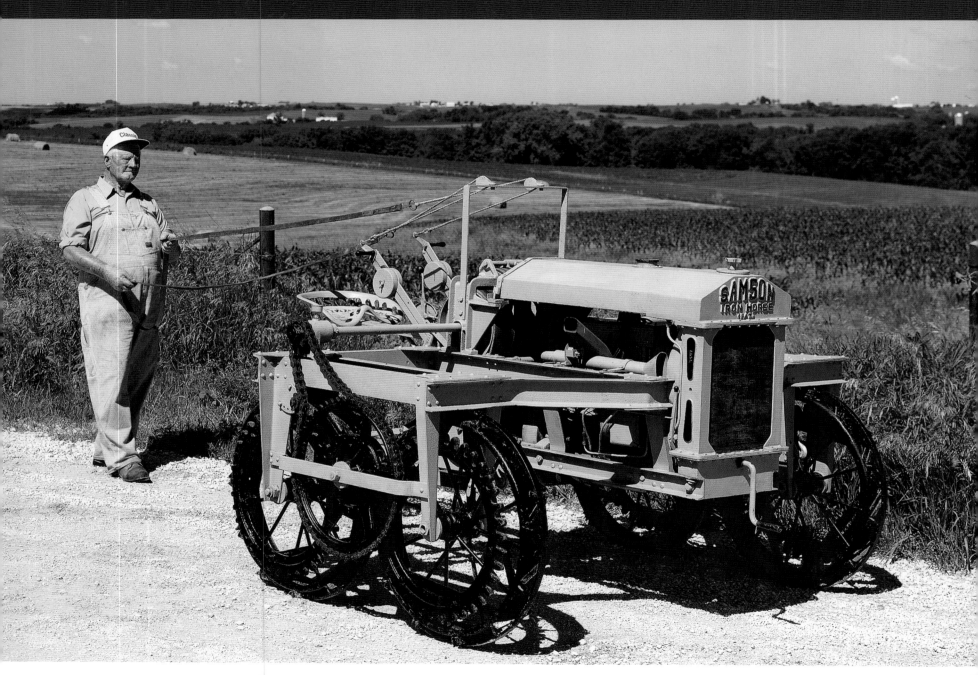

1921 Samson Iron Horse

"Giddy-up!" No tractor so symbolically demonstrates the move from horses to tractors on North American farms as the Samson Iron Horse. In early 1919, GM announced the Iron Horse, a light four-wheel-drive cultivating tractor that could be driven from the seat or with reins, as shown by its collector Eldon Coates of Zwingle, Iowa. GM created the Iron Horse by combining its Chevrolet engine with a cultivating tractor design called the Jim Dandy, which it purchased.

M uscle power—from man, his wife, children, and their
beasts—provided the first power for farming. But an
adult's constant muscle power output is only about one-
eighth horsepower (or 0.1 kW), so the early farmer and
his family looked for help wherever they might find it. Enter the ox and
the horse as prime pullers. And, once the wheel and the lever were figured
out, man began to use those to multiply the amount of power he could produce. One early example of a machine that multiplied the farmer's efforts
is the grain harvester or binder, patented in 1834. It was initially a relatively
simple rig pulled by one horse. But it could do the work formerly done by
five men wielding their scythes and cradles to cut grain. Because of the
harvester's improved productivity, growers suddenly needed a faster way
to thresh their grain. The archaic, but still-used, flail method of beating
the grain out of its husks, or of tramping it out with livestock hooves on the
barn floor, couldn't keep up with the new harvester.

So, the mechanical harvester was soon followed by the threshing
machine. At first, the threshing machine was only a crude coupling of a
pegged spinning cylinder with a grain-cleaning fanning mill. As threshers
grew in size and threshing capacity, and called for even more power to turn
them, sweep "horse-powers" were developed. They harnessed the output
of multiple horse teams. As many as seven teams of horses pulled their
sweeps (levers) around and around together in a big circle to develop the
power needed to run the threshing machines of the day.

Even after combined harvesters were in use in the western United
States, machines cut and threshed the grain in the field where it grew.
Multiple hitches of thirty-five or more horses or mules pulled each big

machine over the vast western wheat fields. Not only were the animals moving the combines over the fields, their power was also driving the threshing equipment and its revolving parts by way of chains (known as link belts) transmitting power from its ground-contacting "bull" or traction wheels.

Though the machine that harvested the grain had evolved, its source of power had not. That horse-powered era of mechanical harvesting, where dozens of horses were harnessed together to pull one machine, is considered the "final development of the heroic age of animal power"—an observation made by Smithsonian Curator of Agriculture John T. Schlebecker, as he collected a 1887 horse-drawn Holt combine for the Smithsonian Institute's Museum of Technology in Washington, D.C.

STEAM AWAITS ITS DAY

By the turn of the twentieth century, steam was quickly becoming the new source of farm power. It held the promise of replacing animal power in agriculture.

Earlier, steam power had played a key role in the development of the early American transportation system. In 1807, Robert Fulton's steam-powered boat, the *Clermont*, began traveling on inland waterways at speeds up to five miles per hour. In 1811, Fulton's *New Orleans* river steamboat traveled the 2,000 miles from Pittsburgh, Pennsylvania, to New Orleans, Louisiana, via the Ohio and Mississippi rivers.

Starting in 1829, the rails of America's early steam railroads were starting to knit the country together. The driving of the

Cradling wheat
Long, steel scythes with wooden "cradles" were used by workers to help harvest up to three acres per day. Helpers following the cradlers gathered the cut wheat into bundles, which were later stacked vertically for further drying before threshing. *State Historical Society of Wisconsin (23435)*

A revolutionary reaper
Cyrus McCormick patented his famous harvester reaper in 1834 and it revolutionized the agricultural world. McCormick's reaper could cut three times as much wheat in a day as a good man with a scythe. It led to other innovations, including the binder and bigger and better threshing machines.

Golden Spike in Promontory, Utah, in 1869 symbolized the tying together of the eastern and western reaches of the United States by rail transportation. These steam-driven machines of travel and transport over water and land encouraged the swift settlement of the West and kicked off an agricultural revolution the likes of which the world had never seen before.

From the time of the California Gold Rush in 1849, steam power's role began to expand as a replacement for human and animal muscle on the farm. The use of steam power on U.S. farms occurred primarily from 1850 to 1925, a period of only seventy-five years. Those years saw farm steam power units advance from small early horse-drawn portable engines like the Archambault small "Forty Niner" (built in Philadelphia, Pennsylvania, in about 1850) to massive 150-horsepower road locomotives, as the engines became more powerful and more useful.

An early market sales leader in the 1840 to 1867 period was George Westinghouse & Company of Schenectady, New York, which produced small portable engines. The Westinghouse upright boiler engine was pulled by a team of horses to its working spot in the farm field or next to the barn. Slowly, steam engines moved away from their stationary status. Some engines were skid mounted and could be dragged between locations. Makers like Westinghouse and Archambault answered the need for engine portability by putting some of their early machines on sturdy wagon wheels. Steam engine maker C. & G. Cooper Company of Mount Vernon, Ohio, started that way, too.

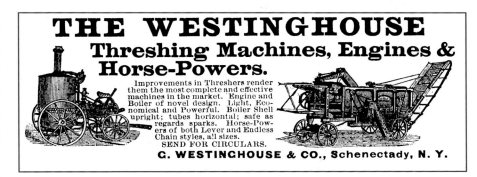

1860s Westinghouse Steam Engine advertisement
George Westinghouse & Company of Schenectady, New York, built pioneering threshing machines, horse-powers, and early steam engines with upright boiler systems. In fact, Westinghouse's steamer was the market sales leader from the 1840s until 1867. Son George Westinghouse Jr. later invented the air brake, and this firm was the forerunner of today's corporation.

1895 Harrison Jumbo 12-HP Steam Engine
The Harrison Machine Works of Belleville, Illinois, developed the Jumbo line of steam traction engines to power the company's threshing machines. This 12-horsepower model, made in 1895, was sold and shipped to William Egbert of Huntingburg, Indiana. In 1930, it was acquired by the Henry Ford Museum in Dearborn, Michigan, as an example of the steam-power era. Bringing the historic tractor full circle, the Ford museum sold it to the city of its creation in 2001. Local volunteers of the Belleville Labor & Industry Museum welcomed it home by restoring it to pristine condition.

EARLIEST FARM TRACTOR

In 1875, the Cooper Company patented a bevel gear drive attachment that converted its engines into self-propelled, or "traction," engines. The unit attached to the engine crankshaft and transmitted power down through a diagonal shaft and back through another bevel gear to the engine's rear "drive" wheels. The steam engine was finally on the move. Threshing's power source could now pull the threshing machine—and its water wagon—from field to field. The Cooper drive unit soon was available to other engine makers, but wasn't built for heavy draft work. Later, sturdier spur gear transmissions filled the bill for the heavy-duty "steam plows" designed and built to plow the prairie.

Although the Cooper firm sold about a thousand traction engines without self-steering, its sales improved appreciably after the steering wheel and gear were added. Cooper quickly became a leading producer of farm engines of the era and made almost 5,000 traction engines in a fifteen-year period from 1875 until about 1891, when it bowed out of the farm steam business.

THE DISADVANTAGES OF STEAM

There were several major strikes against the farm steam-traction engines as they became generally available. First, they were expensive. In 1902, a 15-horsepower J. I. Case traction engine was priced at $1,300, or about three times as much as the threshing machine it powered. At that time, it cost only $50 for a new farm wagon or a two-bottom sulky plow. Often, some form of sharing the cost of a threshing machine and an engine was worked out between farm neighbors. Sometimes that cost-sharing amounted to a partnership between two or more farmers who owned the machines together and then used the machines cooperatively on the farms involved. In other cases, one farmer might purchase the separator and engine and thresh crops for his neighbors on a custom basis, charging for the service by the bushels threshed. Threshing "rings" were the organized local circuit of farmers working together on the harvest. Up to twenty men with their bundle wagons and teams might cooperate in the threshing ring, moving from one farm to the next as the threshing progressed. Tales still abound about the largesse of threshing dinners served to the threshing crew at each farm as the threshing ring made its round.

Another disadvantage to steam engines was that from a cold start each morning, they typically needed an hour or more of firing before they could operate. And to work the engine, you had to "feed" it constantly. Finding fuel and getting it to the engine was a regular chore for the engine operator and his crew.

Best Steam Engine Tractor
Loggers used this behemoth Best steam engine tractor to help pull timber out of California's northern forests. Built by the Daniel Best Steam Traction Company in San Leandro, California, in the early 1900s, steam engine tractors like this were already falling out of favor by the 1920s. When compared to the new gas engine tractors of the day, the steam machines were less efficient, were dangerous, and always needed a large amount of supplied water to run in the fields.

Case Steam Tractors brochure
The dominant steam engine producer, J. I. Case, saw its farm steam engine production peak in 1911 at more than 2,300 engines. By 1915, Case's production numbers were less than half that, and the numbers sold were dropping fast. By 1924, the market for steam power had fizzled out.

Wood was generally available locally, but someone had to cut and split it. Coal, a more dense energy source, had to be brought to the farm from an outside source. Some engines were even designed to burn straw, although it was not a dense energy source.

Another constant need of an operating steam engine was water, and lots of it. A water wagon was part of the rolling equipment that accompanied the traction engine. Depending on how far it was to a water source, one man and a team could be kept pretty busy just pumping and hauling water from local wells or streams to supply a hard-working engine. Steam engines evaporated away about four gallons of water per hour per horsepower. That would amount to about 48 gallons of water used per hour in an engine averaging 12-horsepower output.

The engines were large and heavy, weighing 10 tons or so. And as their power capacity increased, their weight grew, too. Bridges on rural roads built for horse and wagon transport often buckled under a steam engine's weight, and many engineers unexpectedly found themselves in the stream beds below—looking up. Soft ground made by heavy summer rains could easily mire the machines in mud.

Also of concern was the danger of boiler explosions, caused by "crown sheet" failures in farm steam engines. Low boiler water levels or unusual boiler angles on uneven terrain could temporarily drain water from over the firebox and expose the crown sheet above the firebox to a sudden buildup of heat. When water contacted the hot crown sheet as the machine moved back to level, sudden flash steam pressures could blow the boiler apart. Despite the many efforts by engine makers to fix the problem—and the enactment of strict laws made to regulate boiler construction, operation, and maintenance—boiler explosions continued. In 1911, it was estimated there were two boiler explosions every day.

Even with those marks against them, the use of steam traction engines increased slowly from 1880 to 1900. There were nearly 2,000 steam units made per year in the 1890s. That grew to only about 4,000 per year for the next twenty years in the vaunted "golden age of steam."

STEAM THROTTLES DOWN

By the 1920s, the steam age was "losing steam." An alternative source of farm power, the internal-combustion gas farm tractor, was showing its presence. In 1923 alone, more gas tractors were made by one manufacturer—Ford—than all of the steam engines made in the preceding half century. Henry Ford's "Detroit growler" Fordson tractor sold 102,000 units that year.

The dominant traction engine producer, J. I. Case, saw its farm steam engine production peak in 1911 at more than 2,300 engines. By 1915, Case's production numbers were less than half that, and the numbers sold were dropping fast. Case made its last steam engine in 1924.

INTERNAL COMBUSTION INTRODUCED

In 1860, Belgian-born Jean Joseph Etienne Lenoir patented the first commercially produced internal-combustion engine in Paris, France. Its two-stroke cylinder was ignited by a timed electric spark and it burned city (coal) gas. Although it worked better than previous gas engines, its efficiency was poor. Starting from Lenoir's ideas, forty-four-year-old German traveling salesman and mechanical tinkerer, Nicolaus August Otto, developed and, in 1876, patented a four-stroke, or four-cycle, engine. Otto's engine ran more efficiently because it compressed the air-fuel mixture before ignition, thus charging each power stroke with more potential energy. The modern gas engine was born . . . at about the same time the steam farm engine first became self-propelled.

Developments then underway in producing and harnessing electricity were keys to the further refinement of the successful gas engine. Dependable, accurately timed electrical sparks finally succeeded in firing engine cylinder charges precisely, where other methods like igniter tubes, hot bulbs, and flame ignition failed. Otto's engine was soon licensed and sold worldwide. When his patent rights expired about 1890, engine makers mushroomed as they began to use the four-stroke principle for their new compact power source. By 1899 nearly one hundred firms were already making four-stroke gasoline engines.

ROCK OIL WASTE

First called a gas engine (since it burned a gas), Otto's engine arrived on the scene about fifteen years after oil was discovered in Pennsylvania in 1859. That petroleum, or "rock oil," was valued for the kerosene it contained. Kerosene, or "coal oil," burned well in lamps and stoves, but the early refiners were stumped by the smelly gasoline "waste" product that resulted from refining crude oil into kerosene and other fuels and oils. The early refiners often flamed

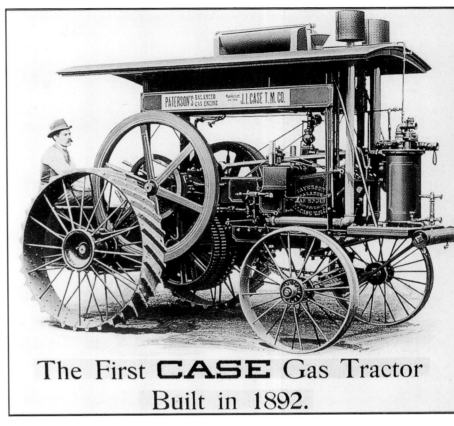

1890s Paterson Balanced Gas Engine
Engineer William Paterson pioneered gas engines for Case in 1892 or 1894. This early machine apparently did not run well enough to be produced, and Case continued to concentrate on building its monstrous steamers until 1911, when it introduced its 30/60 model.

off the volatile gasoline or dumped it just to dispose of it. Otto's engine eventually solved the refiner's problem, as gasoline soon became the preferred fuel for the new internal-combustion engine.

GASOLINE TRACTORS

The first "farm tractor" to burn gasoline as fuel was an early effort by the Charter Gasoline Engine Company of Sterling, Illinois. In 1887, the Illinois firm was at work making single-cylinder, gasoline-fueled stationary engines. Two years later, the company mounted one of its 10-or 20-horsepower engines on a Rumely steam-traction–engine chassis and had an early tractor. Six were put together and shipped to farms in the Northwest. Their fate is unknown.

1892 Froelich Gas Tractor
Thresherman and inventor John Froelich built his own gas-engined tractor in 1892 for threshing work. He mounted a Van Duzen gasoline engine on a Robinson chassis and rigged his own gearing for propulsion. The Froelich, which became the forerunner to the Waterloo Boy tractor, is considered the first successful gas tractor.

FROELICH MAKES IT WORK

A few years later, custom thresherman John Froelich, of northwest Iowa, was apparently searching for a better way to power his J. I. Case 40x58-inch threshing machine. He had previously powered it with a straw-burning steam engine. In 1892, after considering many possible solutions, the inventive Froelich mounted a 20-horsepower, vertical, one-cylinder Van Duzen gasoline engine on a wood-framed Robinson engine chassis. He devised steering and self-propulsion gearing as his creation developed. He ended up that year with a machine that could travel forward and backward. When the threshing season began that summer, Froelich took to the fields with his new contraption. He successfully worked the machine during a fifty-two-day summer-fall season and threshed some 62,000 bushels of small grain.

His simple tractor, a forerunner of the later Waterloo Boy tractor, is generally considered to be the first successful gasoline tractor. The 1892 Froelich tractor "fathered" a long line of stationary gasoline engines and, eventually, the famous John Deere two-cylinder tractor.

Another attempt to make a gasoline tractor that year produced a tractor that didn't quite make the cut. Steam-engine maker J. I. Case recognized the internal combustion engine as a coming competitor to its steam engines. In an effort to develop his own gas tractor, Case hired William Paterson of Stockton, California, to design a two-cylinder, 16/20-horsepower, horizontal-engined gasoline tractor. In his design, Paterson tried to vaporize gasoline for carburetion by bubbling combustion air through a tank of gasoline. Although the machine ran on the gasoline vapors and was tested on belt applications, variations in gasoline volatility and nagging ignition problems caused the machine to be shelved. Paterson's patents on the engine place its design in the 1892 to 1894 period. It was 1911 before Case came up with another tractor.

FIELD READY
FROM HART-PARR

The tractor business really picked up pace in 1896. That year two young men named Charles graduated with degrees in mechanical engineering from the University of Wisconsin: Charles W. Hart and Charles H. Parr. While at the university, they had worked together on an extra-credit project

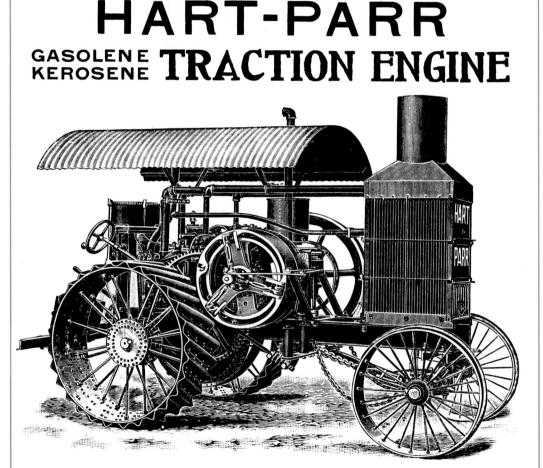

HART-PARR
GASOLENE KEROSENE TRACTION ENGINE

The Great General Purpose Engine

MOST ECONOMICAL ENGINE

For Plowing, Threshing, Road Grading, Shredding, Shelling, Grinding.

HART=PARR CO., Charles City, Iowa.

Hart-Parr advertisement
Hart-Parr of Charles City, Iowa, is the firm credited with coining the term tractor, as well as building the first practical commercially available gas traction engine.

1909 "Minneapolis" Ford
This 1909 Ford tractor was a strange-looking three wheeler that left no "progeny." Hurried into production under the Ford name by Minneapolis, Minnesota, promoters, it had no connection with Henry Ford of Detroit, Michigan. The Minneapolis company incorporated as Ford Tractor Company, using employee Paul B. Ford's name, hoping to trade in on the famous Ford Motor Company reputation. This Ford was powered with a two-cylinder horizontally opposed engine and was driven from the rear steering gear pedestal. Roland Spenst of Alsen, North Dakota, found the pieces of this tractor and put them back together and made it run.

and built several gasoline engines. They were inspired by the potential of their engines and soon found enough financing in the Madison, Wisconsin, area to build a factory to manufacture stationary gasoline engines. In 1901, after making more than 1,000 engines at their Hart-Parr Gasoline Engine Company, they found more financing, moved to Charles City, Iowa, built a factory there, and expanded their firm into gasoline traction engines. Charles City was Hart's hometown.

During the winter of 1900 to 1901, the new Hart-Parr Gasoline Traction Engine Company made its first gas traction engine. Its chassis resembled that of a steam engine, but it was powered by a two-cylinder, 9x13-inch bore-and-stroke engine that produced 17 drawbar and 30 belt horsepower. The gasoline engine ran at 250 rpm and was governed by a hit-and-miss system that produced an irregular exhaust note. Under a load, the engine would fire more often to keep up its speed.

The first Hart-Parr tractor was sold in July 1902 to farmer David Jennings of nearby Clear Lake, Iowa. Hart-Parr's first production run of thirteen traction machines was made in the fall, winter, and spring of 1902 and 1903. In 1903, Hart-Parr enlarged the bore of its engine to 10 inches and thus boosted its output to 22/40 horsepower.

Between the years 1889 and 1901, twelve other companies mounted stationary engines on wheeled chassis arrangements and aimed their machines at becoming traction engines. Some succeeded.

"Ford" Tractor advertisement
An advertisement for the Ford tractor, which was part of the reason the University of Nebraska tractor tests were developed. Nebraska State Legislator Wilmot F. Crozier from Osceola in Polk County was so unhappy with the performance of his "Ford" he spearheaded a law for mandatory tests of tractors sold in Nebraska.

POWERFUL TWINS

From as few as six known tractor makers in 1904, there were as many as 166 manufacturers by 1920, and they were making more than 200,000 tractors a year. At first, these tractors had mostly heavy, single-cylinder stationary-type engines with massive flywheels and were adapted to steam-engine running gears. But soon, twin-cylinder engines were the latest innovation for the new power being made for the farm.

Such "twins" came in many styles and shapes. Some had a horizontally opposed design with a cylinder fore and aft of the crankshaft. Others had both cylinders on the horizontal plane, but they were positioned parallel to each other. Parallel placement of the cylinders on the crankshaft created other design options. Some engines had crankshaft throws placed side-by-side, so they made a power stroke every revolution, like those in the opposed cylinder engines; other designs had the crankshaft throws offset by 180 degrees for better balance. The offset design gave the Waterloo Boy and later John Deere two-cylinder tractors their cheery two-pop rhythm, since the two power strokes came one after the other following only 180 degrees of crank rotation.

McCormick-Deering chose the double-throw crank in its two-cylinder design so that its engine in the popular Titan 10/20 horse-power tractor of 1915 could produce a power stroke every 360 degrees of rotation. The result was a smooth-sounding engine, but with cranks, pistons, and connecting rods changing direction at the same time, the tractor assumed a lively fore-and-aft vibration that could not be completely balanced out at all operating speeds.

Horizontally opposed engines provided inherently superior dynamic balance. Tractor makers Avery of Peoria, Illinois; J. I. Case Threshing Machine Company of Racine, Wisconsin; Pioneer of Winona, Minnesota; Universal of Stillwater, Minnesota; and others opted for horizontally opposed cylinder placement. The big four-cylinder engine in the Pioneer 30/60 of that design ran so smoothly its manufacturer advertised that a nickel placed on its edge on the tractor frame would remain upright while the engine ran.

From hundreds of divergent engine, transmission, chassis, and wheel placement designs, relatively few eventually survived.

TRACTORS AT LAST

In 1907, the Hart-Parr Company began to call its gasoline traction-engines "tractors." The new name was quickly accepted by the industry. Based initially on undercarriages designed for steam engines with huge single-cylinder engines mounted on massive double flywheels, the gas tractors of the age were large, if not always powerful. For the most part, the new gas tractors were intended to step in for tasks steam engines had been used for—powering threshing machines and other belt-driven equipment and doing some heavy tillage duty on large farms. They were still huge, cumbersome, and expensive machines.

However, a few manufacturers came out with machines that moved away from the "heavier is better" trend. Henry J. Heider of Carroll, Iowa, first built a relatively small four-cylinder friction-drive tractor in 1908. By 1911, his Model A machine was on the market. Its four-cylinder Rutenber engine drove the two-speed 12/24 horsepower tractor through a friction plate in contact with the engine flywheel.

A new tractor entry in 1913 was the Bull tractor. It created a lot of interest in smaller tractors and at $335 became a sudden industry sales leader in 1914. The Bull Tractor Company of

1910s Short Turn Tractor
Tractors certainly came in all shapes and sizes in the pioneering years of the 1910s and 1920s. The Short Turn 20/30 was the brainchild of inventor John Dahl and was believed to have been built in both Bemidji and Minneapolis, Minnesota, from 1916 to 1918.

Minneapolis, Minnesota, built the unusual lightweight machine that featured one big "bull" drive wheel on the right or "furrow" side of the tractor. On the left was a smaller "idler" wheel that was not powered. One advantage of the design was that with only one wheel doing the pulling, it didn't need a differential.

Such three-wheel machines abounded between 1913 and 1916. The new "tricycles" included the 1916 Avery 5/10, the 1916 Emerson-Brantingham Model L, the 1914 Farmer Boy, the Rumely All-Purpose, the 1914 Steel King, the 1914 Allis-Chalmers, the 1914 Samson Sieve Grip, the 1915 Happy Farmer, the 1915 Case 10/20, and the ill-fated John Deere Dain All-Wheel-Drive. Like the Bull, some models saw sudden, but fleeting, popularity.

CHOOSING A DECENT TRACTOR

Prior to 1920, tractor manufacturers rated their own tractors as to horsepower and pulling ability, so comparing true tractor specifications was pretty much a guessing game. Field demonstrations, held by organizations such as farm trade magazines, were of some help, but those trials were often subject to individual interpretation. Fortunately for buyers, and ultimately for the manufacturers, objective help was on its way.

In 1918, Nebraska State Legislator Wilmot F. Crozier from Osceola in Polk County was so unhappy with his new tractor's performance that he spearheaded a law for mandatory tests of tractors sold in Nebraska. He was joined in those efforts by fellow State Senator Charles Warner. After their tractor test law passed, the University of Nebraska developed standardized tests that began in 1920 and continue today, more than eighty-seven years later. The tests are used worldwide to ensure that tractor makers don't advertise more performance than their machines can produce. One tractor that spurred the Nebraska tests was a simple three-wheel two-cylinder machine with the "Ford" name. It didn't come from automotive innovator Henry Ford, though. Recognizing the possibility of economic magic in the "Ford" name, promoters from Minneapolis, Minnesota, borrowed the last name of scheme participant Paul B. Ford to form the Ford Tractor Company. Their tractor didn't survive long enough to be tested in Nebraska, but it caused Henry Ford to call his new tractor a Fordson.

THE FORDSON PLOWS NEW GROUND

Henry Ford, of Model T "Tin Lizzie" fame, introduced his revolutionary Fordson tractor in 1917. The standard-tread tractor had its two-piece cast-iron frame components bolted together, making the engine and powertrain into one integral unit. The 20-horsepower four-cylinder gas vertical inline engine ran at 1,000 rpm and was fired by its low-tension flywheel magneto and

Bates Steel Mule advertisement
Call them tracklayers, crawlers, or caterpillars, steel-tracked tractors were created to spread the machine's weight over a larger surface, increase traction, and lessen soil compaction. Made by the Joliet Oil Tractor Company of Joliet, Illinois, the Steel Mule may have lived up to its namesake's proverbial stubbornness as it soon disappeared from the market.

1935 Farmall F-12
The small F-12 Farmall of 1933 was designed as an inexpensive one- to two-plow row-crop tractor. It had most of the operating features of the original Farmall, but was smaller and was priced at $600. Fenders cost $45 as optional equipment. The large-diameter wheels clamped to long single-piece axles and could be adjusted to needed row width. The "tall" wheels gave it the needed under-axle crop clearance. During its production life from 1933 to 1939, 123,442 of the F-12s were made. Rex Miller of Savannah, Missouri, restored this 1935 F-12 soon after he retired from dairying.

coil ignition. All drivetrain components were enclosed within the rear transmission housing. The Fordson weighed just 2,700 pounds.

The thought behind the Fordson's small, cheap design was that it might replace horses and mules rather than steam engines. And it, and the many designs it inspired, got the job done. Despite early sniping from some established farm tractor makers, the Fordson caught on. It was right for its time. And it was priced right, too. In 1918 some 34,167 Fordsons were sold. In 1919, about 57,000 of the gray tractors came off Ford's assembly lines. By 1922, Fordson sales made up 70 percent of all U.S. tractor sales, and in 1923 Fordson sales accounted for a whopping 76 percent of all tractors sold in the United States. Its features and its success soon inspired improved designs and fostered a price war with competing tractor manufacturers.

MORE GROUND GRIP

The success of the Fordson and the lightweight competitors that soon followed caused concern among some tractor makers and farmers. They wondered if drive wheels would lose traction through slippage and if pulling power would decrease too much as tractors got lighter. One idea seen to decrease wheel-slip tendencies was to put more traction surface on the ground with steel tracks, such as those used by Holt and Best, which eventually became Caterpillar. Cletrac (or Cleveland Tractor Company) of Cleveland, Ohio, who made a light track-type machine, went as far as to advertise that their crawler was "Geared to the Ground." In addition to Caterpillar, Cletrac, Monarch, and International Harvester's TracTracTor, other companies made well-known track-type tractors, such as Bates Steel Mule, Bullock Creeping Grip, and Yuba Ball Tread.

The heyday of the little crawlers was relatively short. The additional cost of the tracks and their attendant mechanism, as well as maintenance requirements for the track rollers systems, soon relegated most track-type or crawler tractors to specialized uses in soft soil conditions. As a result, track-type tractors lost sales to more conventional-wheeled tractors. Large, heavy, high-horsepower tractors survived in the Great Plains where their pulling power was needed to work vast acreages of wheat and other small grains. But in the balance of the country, Fordsons and other lightweight tractors soon reigned supreme.

Many four-cylinder vertical inline engine designs were introduced or were already in use by the start of the 1920s. Final drives were, at long last, fully enclosed to spare them the ravages of wear when exposed to the dirt and dust of farm use. Evaporative cooling had disappeared to be replaced by front-mounted radiators using water pumps to circulate the coolant. Effective air cleaners became common to keep the engines from breathing in abrasive dust. Pressure circulation of crankcase oil replaced drip oilers. Friction-prone parts of the engine and transmission were being protected with sealed ball-and-roller bearings. Carburetors had been designed to efficiently burn either kerosene or gasoline. The gas tractor was being refined and defined.

BIRTH OF THE ROW-CROP TRACTOR

With 24 million horses and mules still working on U.S. farms in 1923, and still eating their daily ration of oats and hay, there seemed a large potential market for a tractor specifically designed to replace them. Most of the major farm tractor makers developed and marketed their well-considered solution: a cultivating tractor, a light, single-purpose machine dedicated only to cultivating row crops. Avery of Peoria, Illinois, made and sold four- and six-cylinder cultivating tractors between 1916 and 1919. J. I. Case offered its Cultivator 12 machine in 1919. John Deere experimented with the genre from 1916 to1919, as did Ford. Emerson-Brantingham was in the market by 1923, and so were the Illinois Tractor Company, International Harvester, Parrett Motors, Heider, Ranger, Toro, Universal, and probably others.

Starting with its awkward, and largely unsuccessful, two-row cultivator tractor as an available experimental platform, International Harvester engineers evolved a new type of farm machine, which moved the tractor into another level of farm usefulness: the row-crop tractor. It set a pattern for future tractor design and use. In 1924, IH introduced its Farmall tractor. The light, high-crop–clearance tractor was of a tricycle-gear design. Its rear wheels were set wide apart to straddle two 40- to 42-inch crop rows, and its closely spaced front wheels could travel between those two rows. A crop clearance of 30 inches under the rear tractor axle was achieved by putting a geared drop box from each rear axle. A two- or four-row cultivator could be mounted to the front of the new row-crop tractor. The Farmall was perfect for cultivating corn, a tedious task when using horses.

Instead of the small engines common to the single-purpose motor cultivator, the new IH machine had a four-cylinder vertical inline engine with a 13/20 horsepower rating. The operator's seat was positioned over the tractor's differential with a full view of the cultivator up front. The Farmall was a full two-plow tractor and was equipped with a belt pulley to power stationary equipment and a power takeoff (PTO) shaft to run drawn equipment. It was designed (and thus named) to accomplish *all* of the power needs on the *farm*. At last, farmers could replace their horses and mules with one general-purpose tractor. And they soon did! On the strength of the Farmall's popularity, IH moved ahead of Ford in U.S. tractor sales in 1928.

Spurred by the market success of the Farmall, other tractor manufacturers soon rushed to redesign or create their own smaller tractors with row-crop capabilities. John Deere put together its first row-crop tractor, the GP (or general-purpose) model, in 1928. It had a standard-tread configuration and an arched front axle for clearance to straddle the middle row as it worked three crop rows. Rear-axle crop clearance was achieved with geared drop boxes. Resembling the Model D, the GP was initially a smaller 10/20 horsepower tractor.

The GP also featured a powered mechanical lift to raise and lower its mounted three-row planter and three-row cultivator. Its Power-Lift was an industry first. Next, Deere introduced the GPWT (general-purpose wide-tread) model in 1929. More similar to the Farmall row-crop design, the GPWT was a tricycle-configured machine with a wide rear tread to straddle two crop rows.

Other major tractor makers soon introduced row-crop tractors as they caught on in the late 1920s and early 1930s. In 1930, Allis-Chalmers modified its Model U standard tread into a tricycle row-crop machine and named it the All-Crop Model UC. J. I. Case introduced its tricycle CC general-purpose model. Oliver, recently joined with Hart-Parr, brought out its tricycle row-crop on "tip-toe" rear steel wheels in 1930. The large-diameter rear wheels gave the Oliver tractor the needed under-axle clearance without geared drop boxes. That feature, which reduced the cost of the row-crop machine, spurred other makers to develop second-generation row-crop machines.

Most original of the new designs was one from Massey-Harris. Its 1930 General Purpose tractor was a completely new four-wheel-drive machine. It was available in four different wheel spacings to accommodate working different row widths. Drop gearboxes on all axles gave the tractor the needed under-axle crop clearance. The front wheels steered through universal joints.

RUBBER BOUNCES IN

The usefulness and even productivity of farm tractors was suddenly enhanced in the early 1930s by the switch to pneumatic rubber tires. Today, the use of rubber tires seems a rational improvement, but it was not that easy to see in those economically depressed times. It was only after growers in Florida used truck tires to reduce the damage steel tractor lugs caused on orange tree roots that tractor makers took note.

Working with the Firestone Rubber Company, Allis-Chalmers installed balloon-type tires on their Model U tractor. The first tires installed had no tread and were like the "slicks" used as the main gear tires on the Ford Trimotor airplane, yet their test results were positive. On Labor Day, 1932, near Dodge City, Kansas, a more properly rubber-shod Model U was shown to the public. Soon Goodyear was working with Case to equip its Model C with rubber tires. Not only did the rubber tires take the jolts out of tractor driving, they improved tractor performance, much to the surprise of many observers. Careful scientific studies demonstrated that tractors equipped with rubber tires had better fuel economy at higher operating speeds than those with conventional steel lugs. At the University of Illinois, researchers recorded a 20 to 25 percent gain in useable drawbar horsepower with a 25 percent fuel savings in heavy pulling for a tractor using rubber tires. Also, farmers could save 14 to 17 percent on fuel costs year-round. At the University of Nebraska, researchers found that rubber-tired tractors saved 13.1 percent in time and 17.9 percent in fuel. Their work compared using rubber and steel wheels on tractors for cultivating, combining wheat, binding oats, drilling wheat, picking corn, plowing, and mowing hay.

Although there was initially some resistance by farmers to pay the $200 or so extra for the rubber tires, Iowa State University scientists computed that a tractor had to be used just 500 hours per year to recover that extra cost. By 1933, rubber "air tires" were offered on at least some of the tractors made by the nine leading makers. By 1938, rubber tires were practically considered standard equipment.

1956 John Deere 80 Diesel
John Deere replaced its big standard-tread Model R diesel with the new Model 80 diesel in 1955. The Model 80 diesel was a five-plow tractor with a hefty 61.76 drawbar horsepower and a new six-speed transmission. Its two cylinders had been increased in bore to 6.12 inches and its engine speed boosted to 1,125 rpm to squeeze more horsepower from the design. The Model 80 was built only in 1955 and 1956. It featured the hydraulics and implement mounting available on the other Deere models. This 1956 Model 80 diesel was restored by Mel Humphreys of Trenton, Missouri, with help from friend Chad Reeter.

WHEREVER YOU FARM, IT'S THE SAME STORY:

75% fuel cost savings

MOUNTED TOOL-BAR MIDDLEBUSTERS | MOUNTED TOOL-BAR 2 & 4 ROW CULTIVATORS | ADJUSTABLE WIDE FRONT END | FRONT MOUNTED 4 ROW CULTIVATORS | HYDRAULIC MOWER

SHEPPARD DIESEL SD-3

SOILSAVERS AND PULVERPACTORS

3 POINT AND DRAW BAR PLOWS

SINGLE WHEEL FRONT END

FRONT MOUNTED 2 ROW PLANTERS | ORCHARD AND GROVE TRACTORS | MOUNTED TOOL-BAR 2 & 4 ROW PLANTERS | HYDRAULIC LOADER | FRONT MOUNTED 2 ROW CULTIVATORS

A full line of implements to work any crop with Sheppard Diesel Economy

In practically every state, farmers are using Sheppard Diesels to offset high costs. Their special needs are satisfied by the *wide choice* of Sheppard planters, cultivators and other implements. The big 5 features of Sheppard Diesel power provide *unmatched* versatility . . . put Sheppard Diesels in a class by themselves. For your next tractor, step *up* to the Sheppard Diesel class Mail coupon today. Get all the facts on a Sheppard Diesel.

the only tractor with all big 5 features

½ × ½ = ¼
THE FUEL THE PRICE THE COST

8 SPEED TRANSMISSION

INDEPENDENT POWER TAKE-OFF

BIG DOUBLE DISC BRAKES

INDEPENDENT HYDRAULIC SYSTEM

Sheppard DIESELS
HANOVER, PA.

SEND FOR LATEST INFORMATION ON
☐ Sheppard Diesel Tractors ☐ Cultivators
☐ Planters ☐ Orchard and Grove Tractors
Others
Name
P. O.
R. F. D. _____ State

1950s Sheppard Diesels advertisement
In the early 1950s, diesel power started to reign supreme in the tractor industry. The R. H. Sheppard Company, of Hanover, Pennsylvania, initially offered three rubber-tired diesels, but then expanded its offerings significantly. This advertisement touts one of the main reasons farmers found diesel power attractive: a large reduction in fuel costs.

Rubber shortages during World War II caused the few tractors made in that era to revert temporarily to steel wheels. After the war, rubber tires were the norm again.

BUMPS IN THE ECONOMY

Boom and bust cycles in the U.S. economy and two world wars were important factors in the development and acceptance of the gas tractor. Following the good times just before and after World War I, 186 tractor manufacturers were busy. By 1921, when the economy slumped, the tractors they were making skidded down to 70,000 units. By 1930, only thirty-eight manufacturers were left, but tractor production was up 200,000 machines. Then the situation grew grave again when only 20,000 tractors were made in 1932. By 1933, only nine principal tractor manufacturers remained: International Harvester, John Deere, Case, Massey-Harris, Oliver, Minneapolis-Moline, Allis-Chalmers, Cleveland Tractor, and Caterpillar. Innovations and improvements helped increase tractor sales in the 1930s, as row-crop designs, rubber tires, high-compression engines, streamlining, and new bright colors added to the usefulness, productivity, and even the appeal of the machines. By the time World War II broke out in late 1941, production was back up to 342,093 tractors.

By 1943, the diversion of civilian manufacturing to defense needs had caused tractor production to fall to less then 135,000 units. The recovery after World War II was dramatic. In 1948, nearly 569,000 tractors rolled off the assembly lines. The following year, a total of 141 manufacturers made more than 400,000 tractors. Yet when tractor production caught up with demand, only eight major tractor manufacturers remained in business.

GASOLINE RETURNS

Soon after the first tractors were underway with power from early gasoline engines, economics intervened. Gasoline became expensive as automobile numbers grew and dropped from favor as a tractor fuel. It was replaced by cheaper kerosene and other heavier petroleum distillates.

As early as 1904, Hart-Parr in Charles City, Iowa, invented a new carburetor for its tractors to burn kerosene instead of gasoline. To solve the problem of pre-ignition or engine "knocking" under heavy loads, Hart-Parr added water injection to the carburetor. Then the tractor operator could stop engine knocking by opening a valve to inject water into the engine's combustion air. Hart-Parr's ideas for making kerosene a useable tractor fuel were soon used by other tractor manufacturers, including Rumely of La Porte, Indiana, which used kerosene in its famous Oil Pull tractors.

By 1935, about 95 percent of U.S. farm tractors were burning kerosene and distillates. Typically, those tractors were equipped with small auxiliary gasoline tanks for engine starting. Once started on gas, the tractors were switched over to kerosene or distillate fuel for field operation.

A trend back to using gasoline as a tractor fuel began in 1935. Oliver Farm Equipment Company introduced its new Oliver 70 HC, a row-crop tractor with a high-compression six-cylinder engine designed to burn 70-octane "high test" gasoline.

Added tetraethyl lead, or ethyl, along with better refining methods, improved the performance of gasoline in high-compression engines by raising the fuel's octane rating.

Studies of high-compression tractor engines burning high-octane gasoline at the University of Illinois showed quite an advantage over tractors powered by kerosene. The 1934 trial results showed a 32 percent increase in power at a 12.4 percent fuel savings over kerosene-burning tractors of the same size.

The Oliver 70 HC, with its smooth-running, six-cylinder gas engine, was a hit. Predicting sales of 2,000 units in the new row-crop's first year, Oliver had to gear up to make 5,000 new tractors to meet the market demand. By 1940, nearly a third of all tractors sold from all manufacturers were equipped with new high-compression gasoline engines. The combination of rubber tires and high-compression engines started a trend toward smaller, lighter, more powerful, and more efficient farm tractors going into the 1940s.

DIESELS ADD MORE EFFICIENCY

Caterpillar Tractor Company introduced the first diesel engine in a tractor in 1931 on its Model 65 crawler. International Harvester was next in 1935 when it installed a diesel engine in the Model T-40 crawler TracTracTor. The first diesel-powered wheel tractor came from IH in 1935: the McCormick-Deering WD-40 diesel. Soon diesel engines were common in larger tractors because of their fuel cost efficiencies.

Massey-Harris Four-Wheel-Drive advertisement
In this early Massey-Harris advertisement, the power of four-wheel-drive traction is compared to the traction of old: four horse hooves. Touting its general-purpose four-wheel-drive tractor, the Massey-Harris ad urges farmers to "reduce your operating costs by replacing worn-out machinery and adopting modern methods."

Caterpillar continued to develop new models using larger and larger diesel engines. Cletrac offered a diesel crawler Model DD in 1935. In 1941, the new Farmall Model MD became the first row-crop diesel tractor tested at Nebraska. The first John Deere diesel tractor, the Model R diesel, was first tested in 1949.

As new tractors were developed and marketed, they continued to put horses and mules out to pasture. By 1940, the number of horses and mules on U.S. farms was down to just less than fourteen million, almost half the number reported in 1915, when tractors first started showing up on farms. In 1940, there were 1.567 million tractors reported on U.S. farms by the U.S. Census Bureau.

STREAMLINED TRACTORS

Coming out of the Great Depression, tractor makers not only boosted the performance of their machines but improved their appearance. Industrial design was changing the shape of

1965 Case 130 and 1966 Case 1200

Here, a 1965 Case Model 130 garden tractor sizes up a 1966 Case Model 1200 Traction King four-wheel-drive, highlighting the diversity in size and power tractor manufacturers offered at the time. The 1200 was the first Case four-wheel-drive to enter the horsepower race. The turbocharged 1200 diesel produced 106.86 drawbar horsepower and 119.90 PTO horsepower at 2,000 rpm in 1964 tests at Nebraska. It was the first Case tractor to be turbocharged. Jay Gyger on the Case Model 130 and his father, A. K. Gyger (with brother Kyle on the 1200), are Case experts. The senior Gyger, affectionately known as J. R., is a longtime Case collector. Kyle painted the big Case. The Gyger families live near Lebanon, Indiana.

American products at the time—in everything from telephones to locomotives.

Two of the leading industrial design pioneers applied their talents to reshaping tractors. Henry Dreyfuss, of New York, had completed his streamlining design work on New York Central's *Mercury*, a New York–to-Cleveland passenger train, when he was authorized by Deere & Company to improve the aesthetics of its sharp-edged Model A and Model B tractors in September 1937. By 1938, the John Deere models had been transformed by the new Dreyfuss design. Bold horizontal radiator louvers began at the front of the radiator and swept back to the radiator shroud, making a gentle bend at the radiator edge. The tractor hood tapered back to the operator platform, providing more forward visibility. Russell Flinchum, Dreyfuss' biographer, said the design of the tractor "balances the dominant vertical of the tractor front with flanking horizontal forms to give the illusion of movement when standing still." Flinchum's book, *Henry Dreyfuss, Industrial Designer: The Man in the Brown Suit*, detailed the life and work of the great designer.

International Harvester employed the talents of Raymond Loewy, a New York industrial designer, to give its tractors a new look. Beginning with the IH track-type TracTracTors, Loewy, who helped design the Pennsylvania Railroad's *Broadway Limited*, gave the IH and Farmall tractors a smoothed and rounded sheet metal grille, which gracefully concealed the row-crop tractor's ugly front steering pedestal.

The freshly styled Farmalls introduced in 1939 were the M, H, B, and A models. The newly styled crawlers shown in late 1938 included the big TD-18, as well as the T-6, TD-6, T-9, TD-9, T-14, and TD-14 models.

Loewy also created a new logo for International Harvester, which boldly featured the IH initials, and he later designed a prototype sales and service building for Farmall and International dealerships.

Minneapolis-Moline showed real concern for the tractor operator in its new 1938 designs. The manufacturer's Model UDLX treated drivers better than any prior machine had. The UDLX's fully-enclosed operator cab—complete with lights, starter, radio, spotlight, windshield wipers, heater, horn, and companion seat—was luxurious for a tractor.

The company's big Model U tractor was covered with sweeping lines and fenders that extended from the chromed front bumper to the rear and its folding entry door. The Comfortractor was deluxe

indeed. Unfortunately, it proved to be ahead of its time. The extra cost of its deluxe features detracted buyers and only about 150 were sold. But it is today a prime collector's item.

ANOTHER FORD TRACTOR REVOLUTION

Ten years after Henry Ford's Fordson tractor dropped from view in the United States, the Detroit auto magnate returned to the tractor scene with a new design that once again revolutionized the industry. With an integral hydraulic system working a rear three-point implement hitch on a small tractor, the new Ford Model 9N made history in 1939. The hitch design, crafted by Irish inventor Harry Ferguson, was the key feature of the standard-tread model.

Aimed at smaller farm operations, the Model 9N was an efficient modern-styled two-plow tractor. Its standard wheel tread was adjustable in the front and rear for its use in row-crop operations. The Model 9N Ford-Ferguson sold well—37,283 tractors rolled off factory lines by the end of 1940. Its production was limited during World War II, but it came back strong afterward as the Ford-Ferguson Model 2N.

Even large farms used the handy machine. Easy to operate, the tractor was a machine that kids or grandparents alike could handle. The nimble Ford tractor finally replaced the many mule teams used for light work in the southern United States. Many 9Ns and 2Ns are still at work on U.S. farms, more than sixty years after they were first made.

The design concepts introduced by the 9N had a huge impact on tractors after World War II. Almost every tractor maker eventually adopted integral hydraulics and three-point hitches for rear implement attachment. The Ford-Ferguson design didn't replace the tricycle row-crop machine, but it did become the de facto shape of utility farm tractors.

In 1947, John Deere released its answer to the ubiquitous Ford machine. The Model M offered Quik-Tach implement mounting on an adjustable-tread small tractor (as well as a three-point system similar to the Ford's). The M's all-new engine was still a two-cylinder unit, but was mounted upright inline. In 1948, International Harvester's Model C Farmall became available. Its Touch Control hydraulic system facilitated raising and lowering attached implements, especially the front-mounted cultivator. IH tried a two-point implement-mounting system before finally adopting a three-point system.

The incorporation of integral hydraulic systems and rear-mounted implements was seen in almost all new tractors by the 1950s. Power steering also soon became standard. Because farms were fewer but larger, the demand for more powerful tractors grew. That kicked off a horsepower race among tractor manufacturers. In 1959, John Deere introduced a 215-horsepower four-wheel-drive articulated Model 8010, which upped the ante in the horsepower stakes. Somewhat ahead of its time, the 8010 spurred the imagination of farmers and manufacturers as to what was ahead for farm power.

REPLACING THE LAST HORSE

Tractors aimed at replacing the last horse on farms were often small, one-row machines featuring engines that produced about 10 horsepower. One such machine was the innovative, down-low Allis-Chalmers Model G, introduced in 1948. Its rear-mounted engine pushed the little tractor down the row with the operator seated just above. Its production totaled nearly 30,000 tractors from 1948 to 1955. Another such tractor was International's little Cub Farmall. Like the Model G, it was aimed at the truck farmer. The Cub had an offset-engine drivetrain design with the operator seated above the right-rear axle in the IH Cultivision mode. The Cub was made continuously from 1947 to 1979, thirty-two years, with some 252,997 rolling off company lines. It had the longest production run for any U.S. farm tractor.

THE LEADER

Nearly four million tractors were at work on U.S. farms by 1950, and ten years later only nine major tractor manufacturers remained: Allis-Chalmers, Case, Caterpillar, Ford, IH, Deere, Massey-Ferguson, Minneapolis-Moline, and Oliver.

From 1910 to 1955, International Harvester's tractors captured nearly a third (32.5 percent) of all U.S. wheeled tractor sales. Ford was second with 16.7 percent of sales during those years, and John Deere tractors came in third in sales at 14.5 percent. Ford's sales peaked in the 1920s with its Fordson tractor capturing 44.2 percent of all tractors sold. IH sales bounced back and accounted for 44.3 percent of all tractor sales in the 1930s because of the strength of its Farmall design. From 1910 to 1955, Deere's best sales period was in the 1930s when it captured 21.7 percent of tractor sales, primarily with its Model A and B row-crop machines. Dead last, capturing only 3.1 percent of sales from 1910 to 1955, was Minneapolis-Moline. Oliver was next to last at 4.4 percent of total tractor sales for the period.

John Deere tractor sales finally passed International Harvester in 1963. Deere & Company continues to lead farm equipment sales in the twenty-first century, trying to best hearty competition from manufacturers who have merged over time into ever-changing combinations of ownership.

Initials now only hint at the former manufacturers these new combinations encompass. Case New Holland, or CNH Global, is 91 percent owned by Fiat of Italy. CNH holdings include the historic American farm equipment brands of International Harvester, J. I. Case, Ford, New Holland, and others. AGCO, whose initials once stood for Allis Gleaner Company, has pulled under one umbrella the heritage of Allis-Chalmers, Minneapolis-Moline, Oliver, Massey-Ferguson, and even Caterpillar's Challenger line of agricultural equipment.

THE LEGACY

In the following chapters, the histories of and products built by major farm equipment companies are discussed more in depth, showing how tractors transformed American agriculture in the past century. Everything from a company's beginning and its major product offerings to its survival is detailed. Also told are the stories of some of the machines and their makers that didn't survive the test of time. They, too, are part of the impressive development and productivity of U.S. agriculture and its fascinating history.

To better appreciate the marvelous machines and their poignant history, photographs showing carefully restored examples of these tractors are included. Collected and restored by hundreds of hobbyists, these machines have found homes in many parts of U.S. farm country, stretching from Massachusetts to California and from Texas to North Dakota, and to points in between.

The Farm Tractor Timeline

3,000 BC: Native Americans in today's southwestern states learn to grow maize or corn introduced from South America. Corn culture soon spreads across the continent.

1607: First British colony in America is established at Jamestown, Virginia. The colony goal is to grow tobacco commercially.

1682: Rene Robert Cavelier, sieur de La Salle, claims the entire Mississippi River drainage area for France. To honor King Louis the fourteenth, explorer La Salle names the vast tract Louisiana.

1700s: The Industrial Revolution begins in England and spreads through Europe and North America. Steam power soon augments waterpower and then both animal and human muscle.

1769: Scottish inventor James Watt patents the steam engine. His early pump invention is later modified to produce rotary power through the use of a crankshaft coupled to the piston.

1793: Eli Whitney develops the cotton gin. It opens the eastern United States to textile production. Whitney also pioneers the use of interchangeable parts in making guns. His ideas help establish the factory concept of producing uniform products by machine.

1831: Cyrus Hall McCormick demonstrates his successful grain reaper to his neighbors at Steele's Tavern, in the Shenandoah Valley of Virginia.

1837: Blacksmith John Deere demonstrates a diamond-shaped steel "self-scouring" plow made from a broken saw blade to his neighboring farmers.

FARMALL Plowing and Belt Work Simply Can't Be Surpassed!

THERE is enthusiasm for the work of the FARMALL wherever this perfected tractor appears. On all crops, on all jobs in field and barnyard, it shows the power farmer *something new in handling and efficiency.*

Plowing is one of its strongest suits. The FARMALL owner goes out to tackle that once-dreaded job with interest and good humor. He has learned that FARMALL and its plow will move handily and rapidly over the fields and leave well-turned furrows behind, in ideal shape for the operations and the crops to follow.

On belt work it is the same. We have dozens of positive letters from owners.

D. M. Hastings of Atlanta, Ga., writes, "You deserve a pat on the back for the FARMALL. Please do not thank me for this as it is well deserved." He has used his FARMALL on every kind of work including many belt jobs.

Remember that the Harvester engineers devoted several years to working out this *all-purpose, all-crop, all-year design.* They tried out thoroughly *every* type of design. When FARMALL was *right for all drawbar, belt and power take-off work* they offered it to the farmer. The FARMALL is *the one all-purpose tractor that plants and cultivates, too.* It is the feature of power farming today.

Begin by asking the McCormick-Deering dealer where you can see a FARMALL on the job

INTERNATIONAL HARVESTER COMPANY
of America
(Incorporated)
606 So. Michigan Ave.　　　　　Chicago, Illinois

. . . And next spring your FARMALL will be all ready to go at the PLANTING and CULTIVATING jobs. It's that kind of a tractor!

1841: *Prairie Farmer* magazine, aimed at informing the new settlers about farming the prairie, is founded in Chicago by John S. Wright. He tells his readers about the many improvements coming from emerging farm equipment makers, including Case, Deere, and McCormick.

1842: Young threshing machine operator and salesman J. I. Case arrives in the Wisconsin Territory near Rochester, Wisconsin, from Williamstown, New York, with six "groundhog" threshers. He sells five of the crude machines and keeps one to operate on his own.

1849: Thousands rush to California in search of gold. Stationary steam engines are starting to be used to run mills to saw wood and grind grain.

1861–1865: The bloody American Civil War strains agricultural production and devastates farming in the South. The need for more food and fiber spurs implement makers to improve their machines to further mechanize the farm. Implements designed to be used at faster speeds are made as horses and mules begin to replace oxen as the farmer's prime source of farm power. As a result, farms become more productive with less labor.

1862: The Homestead Act of 1862 is enacted. It provides 160-acre parcels of public land to settlers who erect a dwelling, work the land, and live on the property for five years.

1863: The Holt brothers, Charles, William, Frank, and Benjamin, are attracted to California as a market for the family's eastern-produced finished wood products, including wheels. In 1883, Holt Brothers organize the Stockton Wheel Company in Stockton, California.

1879: Edison perfects the carbon filament lightbulb that soon lights up the world—once electricity becomes available.

1885: Karl-Friedrich Benz of Germany builds the world's first successful gasoline-powered automobile, a single-cylinder three-wheeler.

1892: Custom thresher operator John Froelich of Iowa assembles a one-cylinder gasoline tractor that he successfully uses to power a threshing machine during a fifty-two-day harvest season. His machine can propel itself forward and backward.

1895: Rudolph Diesel of Germany invents the diesel engine, which works with auto-ignition of the highly compressed air-fuel mixture. The engine operates on heavier fuels and is more efficient than gasoline engines, but is more costly to build.

1902: Charles W. Hart and Charles H. Parr, working in Charles City, Iowa, produce their first gas tractor model. The Hart-Parr No.1 is built for farm field use.

1902: International Harvester Company is formed in Chicago by combining the assets of McCormick Harvesting, Deering Harvester, Plano Manufacturing, Milwaukee Harvester, and Wardell, Bushnell & Glessner. The combined companies control more than 80 percent of the reaper, binder, and mower market.

1903: The Wright brothers, Orville and Wilbur of Dayton, Ohio, successfully fly their heavier-than-air Flyer airplane over the sand dunes at Kitty Hawk, North Carolina. They build their own engine, a light, four-cylinder gasoline machine that produces about 12 horsepower.

1903: Henry Ford founds his Ford Motor Company, and five years later his famous Model T is in production.

1904: Benjamin Holt demonstrates his first crawler tractor near Stockton, California. The 1904 tractor is a forerunner of the Caterpillar tractor. Holt is also the leading maker of combined harvesters that both cut and thresh standing wheat.

1914: World War I begins in Europe.

1917: Henry Ford introduces his revolutionary lightweight Fordson tractor and makes it available to England to help it grow food for the war effort. Fordson sales boom after the war's end in 1918.

1918: Deere & Company enters the tractor business when it buys the Waterloo Gas Engine Company of Waterloo, Iowa. The Waterloo Boy two-cylinder 12/25 Model R is then in production.

1920: The U.S. economy retracts after the World War I boom, squeezing tractor makers into mergers or into bankruptcy. The number of U.S. tractor makers plummets from some 186 firms in 1921 to 38 in 1930. Three years later, only nine principal makers survive.

1921: The first commercial radio broadcast is made by station KDKA in Pittsburgh, Pennsylvania. Many other early radio stations are licensed and begin broadcasting that year.

1923: International Harvester announces a revolutionary concept in tractor design. The company's new Farmall is a machine designed for all uses on the farm: plowing, disking, and planting the farmer's crop.

1926: Hybrid seed corn becomes available. Corn hybrids dramatically boost corn yields, and production soars in the 1940s. Soybeans soon join corn as a favored crop in the Corn Belt.

1927: Charles A. Lindbergh captures world attention when he flies his single-engine *Spirit of St. Louis* solo from New York to Paris, France, in 33 hours and 32 minutes.

1929: The Stock Market crash on October 29 puts the U.S. economy in a tailspin from which it doesn't fully recover until World War II. The Great Depression devastates American business, including agriculture.

1931: Caterpillar Tractor Company builds the first diesel farm tractor, the Diesel 65 crawler.

1932: Allis-Chalmers tests the use of pneumatic tires on farm tractors by equipping its Model U with rubber. After trying different air pressures, A-C finds that lower pressures work best and also save traction power.

1936: Electricity is finally available on most U.S. farms as the Rural Electrification Administration (REA) is organized to finance rural power cooperatives.

1939: Henry Ford returns to the farm tractor business with a bang. He introduces the revolutionary, lightweight, two-plow Ford-Ferguson Model 9N with the built-in Ferguson system of implement mounting and hydraulic control.

1941: Liquified petroleum, or LP gas, becomes an alternative fuel to kerosene or gasoline after Minneapolis-Moline engineers modify the carburetion on some of their models to make the clean-burning fuel available for use in farm tractors.

1939–1945: World War II rages in Europe and the Pacific; the United States joins the fight after Japan bombs Peal Harbor, Hawaii, on December 7, 1941. Farm equipment production is greatly curtailed, gasoline and rubber are rationed, and farmers live in a strange new world of price controls in a wartime economy. Farm equipment makers begin producing war materials. Farmers scrap old equipment to be recycled into munitions and war machines.

1945: War in Germany ends on May 7. Two Atomic bombs drop on Japanese cities in August, hurrying the end of the war in the Pacific Theatre and signaling the beginning of the Atomic Age.

1946–1947: Pent-up demand for farm equipment, cars, trucks, and consumer goods keeps manufacturers busy trying to catch up. Steel shortages and labor strikes slow the return of peacetime levels of farm equipment production.

1950: Television reaches the heartland as U.S. TV receivers total 8 million sets. Radio receivers number 45 million.

1952: A small Japanese company called Sony introduces the first pocket-sized transistor radio.

1954: The launch of Russia's Sputnik I, the first manmade satellite, awakens the need for more U.S. technology. The race for space begins.

1958: For the first time, Deere & Company passes rival International Harvester Company in farm equipment sales. To date, IH had been number one.

1959: Commercial jet airline service starts in a New York–Miami route.

1960: John Deere's 42-year-old two-cylinder era ends abruptly August 30, with the unveiling of a revolutionary line of New Generation of Power tractors. Deere engineers invested seven years of secret work on the series to make the 1010, 2010, 3010, and 4010 "magically" appear.

1961: The first factory-installed turbocharger appears on the Allis Chalmers Model D-19. It boosts the power performance of its 262-cubic-inch engine about 25 percent. The gas D-19's drawbar horsepower measures at 63.95.

1965: The International Model 4100 four-wheel-drive tractor comes from the factory with a cab with heating and air conditioning. The six-cylinder turbocharged 429-cubic-inch diesel engine is drawbar tested at 110.82 horsepower. It has eight forward speeds from 2 to 20.25 miles per hour.

1969: The U.S. space program's Apollo 11 lands astronauts Neil Armstrong and Edwin "Buzz" Aldrin on the moon.

1973: The Vietnam War finally winds down with the complete withdrawal of U.S. troops.

1973: International Harvester drops the Farmall name on its tractors some fifty years after the company's row-crop tractors were first given the name.

1977: Horsepower ratings in farm tractors reach their ultimate peak when the 16V Big Bud 747 is custom made in Havre, Montana, by Northern Manufacturing Company. Its big V-16 turbocharged engine operates in the 1,000 horsepower range.

1989: The Berlin Wall falls in Germany. With the collapse of the Soviet empire in 1989, and the end of the Soviet Union in 1991, the Cold War is over.

Chapter Two
Allis-Chalmers

1938 Allis-Chalmers WC
Steel wheels and wide front axles jolted the driver of this 1938 WC. Only a few of the wide front axles were built. The WC was tested again at Nebraska in 1938, after the engine was modified to burn 70-octane gasoline. Belt horsepower went up 39 percent. By the end of WC production in 1948, more than 178,000 had been sold. That compares with total A-C tractor sales of only 28,000 between 1919 and 1933. This 1938 WC is owned by Steve Rosenboom of Pomeroy, Iowa.

The once-giant Allis-Chalmers Company of Milwaukee, Wisconsin, started out small. Unlike many of its later competitors, Allis didn't join the agricultural equipment field until the steam age had nearly ended and internal combustion "gas" tractors had arrived.

Edward P. Allis, a lawyer and businessman from New York, picked up the remnants of a bankrupt milling supply company at a sheriff's sale in Milwaukee, Wisconsin, in May 1861, never realizing tractors were in the company's future. The company was previously owned by Charles Decker and James Seville, who had come to Milwaukee from Ohio in 1847 and established Decker & Seville Reliance Works. The business, advertisements said, was focused on manufacturing French burr millstones, grist, and sawmill supplies. By 1857, Reliance had become the largest iron business in Milwaukee and employed seventy-five men. The Panic of 1857 and its following depression changed all of that and forced Reliance into bankruptcy. Allis gathered up and dusted off the remaining pieces and moved forward.

Allis soon had his new Edward P. Allis Company humming along as he added products to the old Reliance line and added people to his growing customer list. Allis had a knack for figuring out what customers needed, then making those products available. Company sales reached $100,000 in 1865. Two years later, sales rose to $150,000. Recognizing a need for new power sources in milling and related industries—operations once all powered by water wheels—Allis added stationary steam engines to his product line in 1869. Then he bought Bay States Iron Works as he diversified the company. By 1870, sales had grown to $350,000 and he had two hundred employees. When Allis died in 1889, the company already had eleven hundred workers.

The Allis-Chalmers Manufacturing Company came into being in 1901 with the merger of the Edward P. Allis Company with Fraser and Chalmers Company and Gates Iron Works, both of Chicago, Illinois, along with the Dickson Manufacturing Company of Scranton, Pennsylvania. Five years later, A-C headquarters moved back to the Milwaukee, Wisconsin, area from Chicago. That location, now known as West Allis, Wisconsin, eventually grew into a huge industrial complex employing as many as 17,000 people.

Continuing to diversify its product line as it grew, Allis-Chalmers soon became an early industrial conglomerate, long before conglomerates were in. Allis added steam-driven pumps to its product line in the early 1880s and in 1884 built America's largest centrifugal pump. By 1906, A-C was making large steam turbines to power another of its products: electrical generators. After all, the company was in the power and power distribution business.

A FARM TRACTOR

In 1912 a new A-C president, former Wisconsin National Guard Brigadier General Otto H. Falk, was looking for new ways to diversify the company's product line. Internal-combustion "gas" engine farm tractors looked like a good growth area for the company. After first considering producing the Minneapolis-built Bull tractor with its big one-wheel drive, A-C instead chose to design a machine of its own.

In 1914, the new A-C Model 10/18 tractor promised 10 drawbar horsepower and 18 belt horsepower from its 5.25x7-inch bore-and-stroke two-cylinder engine. The tractor utilized a three-wheel arrangement, then common, that placed the tractor's single front wheel in line with the right rear wheel so it would run in the plow furrow. The design was supposed to help guide the tractor while plowing with a moldboard plow. The front wheel could follow the furrow, A-C engineers figured, as horses had done before. That configuration was also used on the Bull tractor that A-C declined to build.

The new tractor's engine turned over at a leisurely 720 rpm, and it had a 303-cubic-inch displacement. The engine had one speed of 2.3 miles per hour forward, and one in reverse at the

1914 Allis-Chalmers 10/18
This 1914 two-cylinder 10/18 Allis-Chalmers tractor was the Wisconsin company's initial entry into the farm equipment business. Its horizontally opposed two-cylinder engine ran at 720 rpm, burning kerosene to produce its 10 drawbar horsepower and 18 belt horsepower. The smaller tank contains gasoline used to start and warm up the engine. This vintage tractor was owned by Richard Sleichter of Riverside, Iowa.

1910s 10/18 advertisement
In this early tractor advertisement, Allis-Chalmers addresses one of the main concerns farmers had about buying a gas tractor: its reliability. This ad touts A-C's 10/18 as a tractor that is "built right and backed right." The ad goes on to note: "It has behind it the reliability that inspires confidence in your trade and insures you against disappointments."

1931 Monarch 50
In 1928, Allis-Chalmers showed an interest in crawler tractors when it bought the Monarch Tractor Company of Springfield, Illinois. This is a 1931 A-C Monarch Model 50 track-type tractor of 53.28 drawbar horsepower. This crawler's power comes from an Allis Chalmers (formerly Stearns) 563-cubic-inch gasoline-fueled four-cylinder engine operating at 1,000 rpm. It was tested at Nebraska at 15,100 pounds. This hefty crawler was restored by Loren L. Miller of Clifton Hills, Missouri.

same speed. Its gear case was enclosed and bathed in oil. Pinion gears from the gear case engaged large ring gears that were attached at the inside of the large drive wheels near their outside rim. The 10/18 was started on gasoline and switched over to kerosene as the engine warmed.

The tractor's weight of 4,800 pounds was light by standards of the day. A-C advertising claimed that it was one of the lightest tractors made. Part of the 10/18's hefty weight came from its one-piece cast-steel frame. It was, said A-C ads, "the only tractor frame with no rivets to work loose—that cannot sag under heaviest strain."

Lacking a distribution network, the A-C 10/18 was initially sold direct from the factory and then shipped by rail to a town near the buyer. The A-C 10/18 was not tested at the Nebraska Tractor Test Laboratory, at the University of Nebraska in Lincoln. The green 10/18 was already on its way out of production when those tests began in March 1920.

A-C made only about 2,700 of the 10/18 tractors during its 1915 to 1920 production life. By then, Allis-Chalmers was already at work designing its replacement machines.

TWO WHEELERS

A-C's second-generation tractor effort was a real "lightweight" at only 2,500 pounds. The 1918 A-C tractor was basically made up of two drive wheels, a transmission, and a steering wheel . . . with a motor. The concept, one already being used by the Moline Plow Company of Moline, Illinois, was that of a two-wheeled power unit that could be hitched to existing horse-drawn machines. Moline called its machine the Moline Universal Model D.

Allis-Chalmers named its new machine the 6/12 General Purpose farm tractor. It had two large front drive wheels with an articulated steering arrangement under which pull-type implements were attached. The tractor operator sat on the seat of the farm implement being used. The LeRoi four-cylinder vertical inline engine was designed to run at 1,200 rpm. The engine had an L-head design with a 3.125x4.5-inch bore and stroke. It had a displacement of 138 cubic inches and one forward speed of 2.5 miles per hour. The final drive was

Allis-Chalmers agricultural yearbook
In one of the company's annual yearbooks, Allis-Chalmers made sure to tout the fact it was the first manufacturer to offer pneumatic air tires on its tractors. The company hired race car driver Barney Oldfield to drive an air-tired Model U at state and county fairs around the country to prove to farmers that the tires worked.

1930 Allis-Chalmers U
First designed to be sold by the United Tractor and Equipment Corporation of Chicago, Illinois, the Allis-Chalmers "United" or Model U arrived in 1929 in Persian Orange paint. The standard-tread machine pioneered the use of pneumatic rubber tires on farm tractors. This 1930 model is equipped with a Continental S-10 L-head engine. In Nebraska tests, the 284-cubic-inch 4.25x5-inch bore-and-stroke engine showed 19.28 drawbar horsepower and 30.27 belt horsepower. This 1930 Allis-Chalmers U is owned by Zachary Lee Schmidt of California, Missouri.

through pinion gears that engaged large drive wheel–mounted ring gears, similar to the final drive of its predecessor, the 10/18. A single-wheel sulky was originally attached to the power unit when the 10/18 was not attached to an implement. A-C later designed a two-wheel sulky for added stability.

Nebraska tests of the 6/12 in 1920 confirmed the A-C power ratings. The Nebraska test board did, however, object to some of the advertising used to sell the machine. The board thought A-C's claim that the 6/12 "stands forth . . . as the most efficient small power unit available" was an excessive and unreasonable claim. Board members also didn't back this A-C claim: "There is no loss of power, no dead weight, no lost motion in the Allis-Chalmers general purpose tractor."

Two or three different versions of the 6/12 General Purpose tractor were made from 1918 to 1926. Most of them were the Model A General Purpose machine. Another version, called the Model B, had a low-set engine for tree clearance in orchard use. A third version, of which only forty-five units were made, was a cane tractor, presumably with high-clearance characteristics for use in southern cane fields. In all, only about 1,500 of the three versions of the 6/12 General Purpose tractor came off factory lines.

1931 Allis-Chalmers UC
Providing row-crop cultivating and farm power utility was the aim of the Allis-Chalmers All-Crop Model UC, introduced in 1930. This 1931 UC row-crop-capable machine is equipped with rubber tires that helped the big tractor roll through the corn field on its tricycle front running gear. The UC used the same Continental S L-head engine used in the standard-tread Model U. This tractor is owned by Chuck Bogaard of Leighton, Iowa.

1936 Allis-Chalmers High Crop
An Allis-Chalmers 1936 Model UC tractor engine and drivetrain were used by Thomson Machinery of Thibodaux, Louisiana, to assemble this A-C Model UC High-Crop tractor. A special bowed wide front end and big rear tires lifted the machine to give it extra under-axle clearance to work in sugar cane and other tall crops. Thomson Machinery used A-C components well into the 1950s to make its specialized tractors. This high-crop machine is owned by Al and Helen Schubert of Republic, Ohio.

The idea of tandem-hitching two 6/12 power units came from A-C's chief consulting engineer, J. F. "Max" Patitz. He optimistically reasoned that if farmers owned two of the small Allis-Chalmers general-purpose units for cultivating and handling light loads, they might like to hitch them together as a four-wheel-drive unit for plowing and disking. One operator with his duplex rig could then pull as many as three plows under favorable conditions, Patitz figured. Although Allis-Chalmers devised a frame with an operator's platform to tandem-hitch two 6/12 units back to back, the duplex concept fell flat and so did the light 6/12 General Purpose tractor.

MAKE IT FOUR

A-C's next approach was making a more conventional four-wheel tractor. Henry Ford's little Fordson four-wheel tractor was selling well, and the A-C engineers took its success into account when they debuted a 1918 Model 15/30, a three-plow conventional tractor that was bigger than the Ford's two-plow size. When Nebraska tests showed it produced more than its rated 20 drawbar horsepower and 30 belt horsepower, A-C renamed the machine as its 18/30 model.

Allis engineers developed their own engine for the new 18/30. It was an inline-vertical four-cylinder mill with a 461-cubic-inch displacement coming from its 4.75x6.5-inch cylinders. With the two-speed transmission and the engine turning at 830 rpm, the tractor moved forward at 2.31 miles per hour in first gear and 2.82 miles per hour in second. It was heavy at 6,000 pounds—more than twice the weight of the Fordson (2,700 pounds).

The Allis-Chalmers standard-tread tractor became even heavier in 1921, weighing in at 6,640 pounds, as it entered a new phase in its life. At its Nebraska tests in 1921, the new version of the 18/30 showed enough horsepower to rename it the 20/35. It produced 25.45 drawbar horsepower and 43.73 belt horsepower in those tests. The 20/35 burned gasoline instead of kerosene, and the engine rpm had been stepped up to 930.

NEW A-C TRACTOR DIVISION

Figuring tractors were here to stay, Allis-Chalmers created a new tractor division in 1926, which had its own distribution and sales system. Along with launching the new division in 1927, A-C completed a major redesign of the 20/35 tractor, giving it a new look.

1936 Allis-Chalmers U on rubber
This 1936 Model U was stronger and faster than its earlier stablemates. It could handle three 14-inch plows at 5 miles per hour, Allis claimed. The 1936 U also had a valve-in-head 318-cubic-inch UM engine with a 4.5x5-inch bore and stroke. It could race down the road at 11.75 miles per hour. Changes on the Model U included adding larger-diameter rims and tires, a redesigned platform, and wider fenders. A new padded seat for the U's industrial version, the UI, aided operator comfort. This 1936 Model U on rubber tires was collected by Tom Luetkemeyer of Belleville, Illinois.

When tested at Nebraska in 1928, it produced 33.20 drawbar horsepower and 44.29 belt horsepower. The engine was of the same specifications as before. The added drawbar pull came from a heavier working weight: 7,095 pounds. The extra weight reduced the tractor's wheel slip. By 1929, the old 20/35 became known as the E20/35. In following years, it was designated the Allis-Chalmers Model E.

THE L, YOU SAY

Allis-Chalmers needed a somewhat smaller tractor than the 18/30 platform, so in 1920 it began making the 12/20, destined to become the Model L. It too had a four-wheel standard-tread design. Allis powered the 12/20 with a four-cylinder valve-in-head engine it bought from Midwest of Indianapolis, Indiana. The Midwest engine got 280.6 cubic inches of displacement from its four 4.125x5.25-inch cans. It turned at 1,100 rpm.

First rated as 12/20 horsepower, the tractor was tested at 21.42 drawbar horsepower and 33.18 belt horsepower at Nebraska and was immediately renamed the 15/25 model. It had a two-speed transmission, which was capable of reaching 2.3 and 3.1 mile-per-hour speeds. With good sales figures to start, the 15/25 eventually did not prove to be as popular as the larger tractor. In 1927, the 15/25 was discontinued. Between it and the Model L 12/20, 1,708 copies came off the factory lines between 1920 and 1927.

TRACTOR DIVISION GROWTH

Starting from zero in 1914, Allis-Chalmers progressed through three basic tractor designs and was producing tractors that had gained acceptance among customers by 1937. Production of the A-C models 15/30, 18/30, 20/35, and 25/40 had reached 16,425 units and firmly established A-C as a tractor manufacturer. Even better times were ahead.

CRAWLER TRACTORS

Showing once again its penchant for diversification, Allis-Chalmers added to its tractor division in 1928 when it bought the struggling Monarch Tractor Company of Springfield, Illinois. Monarch made a line of track-type tractors, including a mammoth

11.5-ton Model 75. Monarch first opened shop in Watertown, Wisconsin, in 1914 and began making its crawler tractors there in 1917. By 1928, Monarch was located in Springfield and was in financial distress when A-C purchased the crawler tractor company plant and assets for a half million dollars.

Allis-Chalmers soon stopped production at Monarch while it redesigned the Monarch line. First presented as a new A-C crawler was the Model 35, a 41-drawbar horsepower machine. A new Model L crawler replaced the big Monarch 75. A Model K soon replaced the Model 35 and Allis quit using the Monarch name on the machines. The crawlers were now Allis-Chalmers tractors.

A new farm-size model, the M, featuring new steering clutches was a 35.5-drawbar horsepower machine that burned

1931 Rumely Six A
Short-lived six-cylinder tractor power from Advance-Rumely came to Allis-Chalmers in 1931 when A-C bought the old full-line thresher-tractor company. This 1931 Rumely Model Six A was developed to modernize Rumely's aging line of OilPull tractors. The four-plow tractor showed 33.57 drawbar horsepower and 48.7 belt horsepower in Nebraska tests. Since it competed with the Allis Model E, the Rumely Model Six A was not manufactured by Allis after its inventory from Rumely was gone. This Model Six A was collected by Norman Steinman of Bourbon, Indiana.

1913 Rumely E OilPull

Long a mainstay in the Advance-Rumely line of OilPull tractors, this 1913 Model E, or 30/60, shakes the ground with its 13-ton weight when it ambles by. The big tractor was made from 1910 through 1923. Two 10x12-inch bore-and-stroke cylinders burning kerosene turned the engine at 375 rpm to give it 50 drawbar horsepower. To control pre-ignition knocking, water was injected into the engine's combustion chamber. Oil was the coolant in the big front-mounted radiator. Wide extension rims kept the big tractor on top of soft ground. Ron and Lora Lea Miller of Geneseo, Illinois, restored this big OilPull.

either gasoline or distillate. The Model M was also available as an orchard model with track-shielding fenders and the operator's seat placed low behind the tractor for limb clearance. A-C continued making the farm-size crawlers at Springfield, but was anxious to expand into larger industrial machines.

The big diesel-powered industrial series of A-C crawlers started production in 1939 with the HD14 model. It was powered by a General Motors two-cycle diesel. By World War II, Allis was also making GM-powered HD7 and HD10 two-cycle diesel tractors. After the war, the HD5 and HD19, at the bottom and top ends of the power range, were also manufactured in Springfield with GM engines supplying the diesel power.

1923 Rumely G OilPull
This 1923 Model G 20/40 tractor was made by Advance-Rumely from 1919 to 1924. Its two-cylinder 8x10-inch bore-and-stroke kerosene engine powered it up to 30.7 drawbar horsepower and 46.19 belt horsepower in 1920 Nebraska tests. Rumely prided itself in producing engines that exceeded its own horsepower ratings. Like the other OilPull models, the engine cooling draft was created by engine exhaust blowing through the radiator. Rumely-lover Scott L. Thompson of Tremont, Illinois, collected this Model G.

PERSIAN ORANGE MODELS

Some years before most U.S. tractor makers brightened their machines with new brilliant colors, Allis-Chalmers developed a landmark tractor painted bright orange. In 1928, United Tractor and Equipment Corporation of Chicago asked A-C to manufacture a new 4,000-pound tractor to replace the Fordson it was then selling. United was a group of thirty-two independent farm and industrial equipment makers just getting organized to sell tractors and equipment, and A-C was a member of the group.

A-C agreed to make the tractor, selling it through the United organization (the reason why the new model was called the "United" tractor). It was basically a redesigned Model 20/35. The earlier Allis-Chalmers Model 20/35 had been discontinued the year before so the new tractor also filled a horsepower gap in A-C's lineup.

The new United tractor model was launched at the 1929 Southwest Tractor and Road Show in Wichita, Kansas. It later became known as the Model U when United Tractor & Equipment Corporation dissolved. Apparently, the bright orange paint was a marketing idea of the United group. Allis continued the use of the color and soon all A-C tractors were issued in Persian Orange paint. Harry Hoffman, A-C tractor advertising manager at the time, later said that was how A-C "chose" its orange color.

Continental of Muskegon, Michigan, supplied the S-10 flathead engine for the original United tractor. The Continental S-10 was a four-cylinder 4.25x5-inch bore-and-stroke engine with 284 cubic inches of displacement. The engine spun at 1,200 rpm and gave the 4,281-pound tractor 19.28 drawbar horsepower and 30.27 belt horsepower when it was tested in Nebraska in 1929. The first 7,404 Model Us built from 1929 to 1932 used the S-10 Continental engines. A-C equipped later Model U tractors with a Waukesha UM valve-in-head engine. The UM was a 4.375x5-inch bore-and-stroke engine with 300.7 cubic inches of displacement. In 1936, the UM engine upped its displacement to 318 cubic inches when its bore was increased to 4.5 inches.

RUNNING ON AIR

The Model U soon gained another historic distinction besides being the first orange-colored tractor. Allis-Chalmers and its Model U pioneered the use of pneumatic rubber tires on farm tractors. A-C claimed the Model U was the first farm tractor to be equipped with pneumatic tires by a tractor manufacturer.

Moving from a wild-sounding idea to field trials took a relatively short period of time. Harry Merritt, the general manager of the Allis-Chalmers tractor division, talked over his idea to use pneumatic tires on the Model U with his engineering heads before calling Firestone Tire and Rubber in Akron, Ohio, for its input. When Firestone got involved, it furnished a pair of smooth (non-treaded) airplane tires that were soon fitted to the Model U.

Early tests showed that by reducing the tire pressure from 70 down to about 12 pounds per square inch, the tires worked well. Further tests produced even more encouraging results. Once the major problems were solved, Allis-Chalmers showed and demonstrated the Model U on rubber tires near Dodge City, Kansas, on Labor Day in 1932. On its new rubber tires, the U had working speeds of 2.5, 3.33, and 5 miles per hour in the field. Its fourth gear, which was locked out on steel-wheeled tractors, let the rubber-tired U motor down the highway at speeds up to 15 miles per hour.

But offering rubber tires on tractors and selling them to farmers were separate matters. Allis-Chalmers had to market the new tire-equipped tractors with demonstrations and even rubber-tired tractor races at farm fairs. Popular Indianapolis auto racing champion Barney Oldfield was often featured as the winning driver on a souped-up Model U. Oldfield made one run in September 1933 that clocked out at 64.28 miles per hour while driving the rubber-tired U. Previously, he had been the first race car driver to exceed 60 miles per hour in an automobile.

Rubber tires soon became popular, and by 1938 they were practically standard equipment on farm tractors. Part of the reason was that tractors equipped with tires could do their work faster. The rubber-tired Allis Chalmers Model U could pull three 14-inch plows at 5 miles per hour. When compared with steel-wheeled tractors, which could only reach 3.5 miles per hour with the same plow, that was an impressive achievement. In the end, Allis built more than 23,056 Model U tractors from 1930 to 1953.

A ROW-CROP TRACTOR FOR A-C

In a measured response to the success of row-crop tractors, in particular the Farmall, Allis-Chalmers introduced its All-Crop Model UC in 1930. The UC was a high-clearance, tricycle-type row-crop machine based on the Model U. It used offset final-drive gearing to boost the rear-axle clearance for late cultivation of row crops.

1916 Rumely F OilPull
One 10x12-inch cylinder kept this 1916 Model F 15/30 OilPull underway. An extra large flywheel helped smooth the power strokes of the engine turning over at only 260 rpm. Rumely advertised this tractor as having the power of fifteen good draft horses and the endurance of fifty at a cost of less than ten. This tractor is one of about thirty different models made by Rumely between 1910 and 1931. Total Rumely production for that period was 56,647 tractors. Paul Stoltzfoos of Eola, Pennsylvania, collected this 1916 Rumely F OilPull.

Like the Model U, the UC was originally powered by the Continental 284-cubic-inch engine. In 1933, the engine on the UC was upgraded to the UM 300.7-cubic-inch engine. In 1936, its power was bumped up again with a 318-cubic-inch engine. A few more than 5,000 UCs were made between 1930 and 1941, and Thomson Machinery of Thibodaux, Louisiana, used UC components for many years to make its high-clearance cane tractors.

WELCOME OILPULLS

Allis-Chalmers took over the Advance-Rumely Thresher Company of La Porte, Indiana, in June 1931. Rumely was the manufacturer of the famous long line of OilPull tractors. With the purchase of Advance-Rumely came its La Porte plant, some branch houses, the Advance-Rumely line of equipment, and dealerships with their valuable customer base.

The old full-line company had a long history, going back to the mid-1850s when brothers Meinrad and John Rumely's company, M. & J. Rumely, and their preceding firms, pioneered threshing machines, stationary steam engines, and traction engines, as well as the famous line of Rumely OilPull oil and gas tractors. Advance-Rumely was the combination of previous Rumely organizations and was organized in 1915. The following businesses were part of the consolidation: Gaar-Scott & Company of Richmond, Indiana, dating back to 1836; Advance Thresher Company of Battle Creek, Michigan, founded in 1881; and Northwest Thresher Company of Stillwater, Minnesota, dating to 1884. In 1924, Advance-Rumely bought Aultman & Taylor Machinery Company of Mansfield, Ohio, another longtime steam-engine and threshing equipment maker that was founded before the Civil War.

The first OilPull came off the Rumely production lines in early 1910. By October that year, 100 had been made. Rumely produced a total of 56,647 OilPulls in fourteen different models by the time the company was bought by Allis-Chalmers. The OilPulls were sturdy machines whose engines were cooled by massive box-shaped oil-filled radiators. They drew their radiator cooling air by inducing a draft of engine exhaust through them.

The carburetor development on the OilPulls came from John A. Secor, an early New York experimenter who created a way to use low-cost fuels for marine internal-combustion engines. Secor joined the Rumely company in 1908 as chief engineer. Secor and his nephew, William H. C. Higgins, are credited with developing

the successful kerosene-burning carburetor used on the OilPulls. Their carburetor atomized kerosene and relied on water injection to keep the big cylinders from pre-igniting and causing power-robbing "knocking" under heavy engine loads. Secor held nearly twenty U.S. patents relating to engines.

DOALLS, TOO

Advance-Rumely's rather late efforts to make and sell a general-purpose or row-crop tractor centered on the DoAll tractor. Rumely bought the DoAll design in 1927 from Toro Manufacturing Company of Minneapolis, Minnesota.

1921 Rumely 12/25 OilPull
Somewhat smaller than the other OilPull models was this 12/25, made in 1921. The OilPulls were widely admired for their steady power, fuel economy, and longevity. Rumely downsized its line of OilPulls starting in 1924 and continued their production until 1930. As the threshing machine gave way to the combine in the late 1920s, the market for the OilPull tractor waned. Soon lightweight tractors from Allis-Chalmers, International Harvester, and John Deere were all the rage. This 12/25 OilPull is owned by Kent Kaster of Shelbyville, Indiana.

The DoAll was an early attempt at a general-purpose tractor that combined features of a cultivating tractor with those of a conventional tractor. Two Rumely DoAll models were available. One model converted to a motor cultivator by shifting the rear wheels forward and then removing the tractor's front axle and wheels. A large caster tail wheel attached to the cultivator, then supported the rear of the tractor. Conversion time from tractor to tractor-cultivator was estimated at a half a day. The lightweight 20-horsepower machine was also available without the conversion feature. More than 3,000 DoAlls were built by Rumely. A-C dropped the DoAll design. It had a better row-crop tractor.

RIGHT TRACTOR, RIGHT TIME, RIGHT PRICE

The second-generation Allis-Chalmers row-crop tractor was exactly what farmers needed when it was introduced in 1933, during the depth of the Great Depression. The new two-plow Model

1915 Aultman-Taylor 30/60
Aultman-Taylor of Mansfield, Ohio, a heavyweight power contender and constant rival of Advance-Rumely, was the builder of this big machine. Rumely bought the Ohio company in 1924 and closed it down. This 1915 Aultman-Taylor Model 30/60 was a favorite of rural road districts, where it was used to grade country roads. The four 7x9-inch cylinders in its transverse-mounted engine ran it at 500 rpm. Its 120-gallon water radiator used dual fans to pull cooling air through its 196 two-inch tubes. The gas tractor weighed about 11 tons. Mel and Lois Winter of Minneota, Minnesota, collected this mighty machine.

WC row-crop was the first tractor to be offered with pneumatic rubber tires as standard equipment. The WC's base price was $825. It cost $150 less, or $675, if a farmer decided to order the WC with optional steel lugs.

Built primarily from off-the-shelf components to pare manufacturing costs, the tractor had an automotive-style transmission coupled to a truck-type differential. A rugged, steel-channel frame kept it lightweight without the need for the heavy castings used in the U and UC (the castings compensated for their frameless construction).

The twenty-five tractors made in 1933 were equipped with 186-cubic-inch engines from Waukesha, which had bores and strokes measuring 3.625x4.5 inches. The now-famous Allis "W,"

a four-cylinder 201-cubic-inch engine, was installed in the WC models beginning in early 1934.

The WC had the distinction of being the first tractor tested on rubber tires at the Nebraska tractor test station. The rubber-tired WC turned out 19.17 drawbar horsepower versus only 14.36 horsepower from the same tractor on steel. Belt horsepower was 21.48 horsepower.

The WC had a four-speed transmission, giving it working speeds of 2.5, 3.5, 4.75, and 9 miles per hour. An optional engine-driven power lift, which made its appearance on models produced after 1934, helped take some of the repetitive lifting work out of cultivating row crops by using the front-mounted two-row cultivator.

1937 Allis-Chalmers A

This Allis-Chalmers Model A of 1937 promised to bring big power to the farm. It replaced the A-C Model E in 1936. The A was rated as a 33/44 machine, capable of pulling four plows or running a big thresher. Its four-cylinder 460-cubic-inch gas engine had a 4.75x6.5-inch bore and stroke and was designed to run at 1,000 rpm. The distillate version featured more displacement at 610 cubic inches. The A's four-speed transmission was a vast improvement over the two speeds in the Model E. A little late as a threshing tractor, the Model A only sold 1,225 units before it was discontinued in 1943. This Model A is owned by Dick Heitz of Delphos, Ohio.

STREAMLINING

In 1938, the WC joined the growing group of farm machines that were showing new stylish lines. Rounded clamshell fenders, rounded radiator grilles, and rounded-off gas tanks gave the old Depression-era workhorse a younger look. Industrial designer Brooks Stevens of Milwaukee worked the metal magic. The WC was also modernized with a new electrical system that added an optional starter and electric lights.

The redesigned WC also had increased power. Belt horsepower was up to 25.45 with 20.41 horsepower delivered in pulling power at the drawbar. In 1938, a high-compression gas-engined WC was tested, and it put out 29.93 belt horsepower and 24.16 drawbar horsepower. The WC was redesigned and refined.

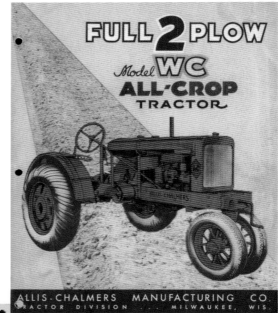

1930s WC All-Crop brochure
In the brochure introducing Allis-Chalmers' WC all-crop tractor, the company emphasized that the tractor could pull two full plows. In fact, on rubber tires, the WC could pull two 14-inch plows at 4.75 miles per hour.

1933 Allis-Chalmers WC
First of the famous Allis WC row-crop model line was this 1933 model, known as the W All Crop. Designed to use pneumatic rubber tires, the WC led the field as a low-cost row-crop tractor. Built on a simple steel channel frame, the WC used off-the-shelf drivetrain components to pare costs. The Waukesha 186-cubic-inch 3.625x4.5-inch bore-and-stroke engine was used in the first twenty-five tractors. The WC's powerplant was upgraded in 1934 to the A-C 201-cubic-inch 4x4-inch bore-and-stroke engine. Laverne Greif of Dallas Center, Iowa, collected this row-crop tractor.

1937 Allis-Chalmers WC
Hints of streamlining shape this 1937 Allis WC. The formerly all-flat radiator grille is slightly bowed forward at the bottom, and a graceful curve shows in the tractor hood. This tractor has the optional mechanical power lift for raising the two-row cultivator. Two pins attach the drive-in two-row cultivator to the two front-mounting brackets on the tractor frame above the front wheels. Operator comfort was addressed with a back support on the steel pan seat. Kenny Kamper of Freeburg, Illinois, collected this updated version of the WC.

1935 Allis-Chalmers WC
This 1935 A-C Model WC rides on rubber tires and rims, converted from original steel wheels. The first tractor tested on rubber tires at the Nebraska Tractor Test Station, the WC proved a 33.5 percent horsepower advantage over the same tractor tested on steel wheels. The tire-equipped WC put out 19.17 drawbar horsepower versus only a 14.36 horsepower rating from a steel-wheeled WC. Rubber tires, despite their extra cost, soon became almost standard. This WC tractor was restored by Dewey McIlrath of Rossville, Indiana.

The tractor proved to be a winner for Allis-Chalmers. During its production run from 1934 to 1948, some 178,000 WCs were made. Its peak production numbers came in 1937 when 29,208 units were manufactured. Its popularity with farmers boosted A-C to the third best-selling tractor maker in the United States in the 1930s, capturing 12.6 percent of all wheel tractor sales.

GIVE ME A BIG A

The next tractor Allis-Chalmers introduced was a four-plow tractor in a standard tread arrangement and was targeted to larger tractor users. The hefty Model A was unveiled in 1935 and it replaced the Model E. The powerful Model A, with its four-cylinder 4.75x6.5-inch bore-and-stroke engine had a displacement of 520 cubic inches. It turned out 44 belt horsepower.

The Model A was a heavy tillage and belt power specialist. It improved on its preceding model with a four-speed transmission that was capable of 2.5, 3.75, 5, and 9 mile-per-hour forward speeds. Only about 1,200 Model As were built and sold by Allis during its production run from 1935 to 1943. With the advent of the combine, such heavy "threshing" tractors were no longer in demand.

GIVE ME A LITTLE B

After the mighty Model A, A-C decided to scale its next tractor down a bit. The company designed a one-row, one-plow, "trade in

1938 Allis-Chalmers B

"For all jobs on small farms, for small jobs on big farms," read Allis-Chalmers' ads for its new Model B tractor, announced in 1937. Designed to replace the farm's last team, the early Model B put out 12.97 drawbar horsepower from its 116-cubic-inch distillate-burning engine. The B's thin torque tube wasp-waist frame started a trend in small tractors. A-C let the operator decide which side of the row to watch by providing a wide bench cushion seat and wide backrest. This 1938 Model B, owned by Dwight Hart of Jackson, Tennessee, wears factory steel wheels.

the last team of horses" tractor aimed at small family farms and truck patch growers. The Model B was introduced in early 1938 with a bowed front axle and wide-set front wheels designed to straddle the one row it could cultivate. Offset rear-wheel drives gave the Model B high under-axle crop clearance.

A small-diameter torque tube frame allowed for extra visibility from the operator's platform to the crop rows below. That construction was first featured on the Model B, but was soon in use on other similar small tractors. Instead of offsetting the driver's seat to one side to see around the engine, as other similar tractors had, the

A-C design had a wide-cushioned bench seat on the B, allowing the driver to slide to the best spot to see the cultivator.

Industrial designer Brooks Stevens again worked his design tricks on the Model B. He rounded the front and corners of the radiator and the fuel tank, and designed a sculpted clamshell fender. They were all touches that gave the tractor a solid but modern and friendly look.

The first ninety-six Model Bs were powered by a 3x4-inch bore-and-stroke Waukesha engine. In 1938, a new Allis-Chalmers 3.25x4-inch bore-and-stroke engine with 116 cubic inches of displacement was standard in the B. It produced 15.68 belt horsepower and 12.97 drawbar horsepower on distillate at 1,400 rpm in Nebraska tests. The engine was later tweaked to a 125.2-cubic-inch-displacement gas-burning unit, which put out 22.25 belt horsepower and 19.51 drawbar horsepower. The power upgrades came as a result of a bore increase to 3.75 inches and an engine speed boost to 1,500 rpm.

The Model B sold for $570 in early 1941 when equipped with rubber tires, lights, a starter, a muffler, and a radiator shutter. On rubber, but without the other amenities, it listed at $495.

The Model B was a welcome power addition to many farms. Allis-Chalmers sold more than 118,000 Model B tractors from 1938 to 1958. An industrial version of the B, the Model IB, was also offered, and 2,800 were built.

TWO-ROW MODEL C

A derivative of the Model B was the light two-plow, two-row Model C, which Allis debuted in 1940. Available with either a narrow tricycle front or adjustable wide front, the Model C employed the same 125.2-cubic-inch Allis engine that was used in the B. Its torque tube "wasp-waist" construction was also similar to the B's.

Burning gasoline, the Model C put out 23.3 belt horsepower and 18.43 drawbar horsepower. Distillate burners tested at slightly lower outputs. The narrow, dual-tire-front version was the most popular of the front-wheel options. Slightly more than 84,000 Model Cs were built between 1940 and 1950. If equipped with rubber tires, lights, a starter, and a muffler, the Model C sold for $595 in early 1941, just $25 more than the Model B.

INTERIM MODEL RC

A relatively rare Allis, the RC, was made and sold from 1939 to 1941, during the time the Model C was being developed. Basically the RC was the WC tractor chassis powered by the Model B's 125-cubic-inch engine. By sharing the WC frame, it could mount and use equipment made for the WC.

Allis production figures indicate 5,516 of the interim design Model RC were built. Its four-speed transmission gave it working speeds of 2, 2.8, and 3.75 miles per hour with a transport speed of 7.5 miles per hour in fourth gear. Rated as a light two-plow tractor, it was only 500 pounds lighter than the WC.

Limits on manufacturing and rubber rationing during World War II reduced the production of A-C tractors. Many that were made were factory-supplied on steel wheels. After World War II, product improvements multiplied. Chief among post-war developments at Allis-Chalmers was the launch of the new Model WD tractor and an unusual, but prolific, "one-horse" machine from down South.

LOW-DOWN "WHEELED HOE" G

A unique Allis-Chalmers tractor, the Model G, was introduced in 1948. Designed for the truck farmer who might still have one horse yet to replace, the low-profile machine seated its operator only inches above the one crop row it was cultivating. A rear-mounted 10-horsepower four-cylinder 62-cubic-inch Continental engine pushed the 1,285-pound "hoe on wheels" down the row at speeds as slow as three fourths of a mile per hour. The Model G was ideal for cultivating tender young plants.

The Model G's Continental engine had 2.375x3.50-inch bore and stroke and revolved at 1,800 rpm. The engine had a compression ratio of 5.75:1 and burned gasoline. The small machine developed 10.33 belt horsepower and 9.04 drawbar horsepower. The four-speed transmission gave it the right speed for the job at hand. The G's twin-tubular front steel frame allowed plenty of visibility and enough under-frame clearance to mount a special 12-inch moldboard plow, a one-row cultivator, or other job-specific implements.

The tractor was developed in West Allis, Wisconsin, but built in an Allis-Chalmers plant in Gadsden, Alabama, where 29,036 Model G units rolled off the assembly line from 1948 to 1955.

WD SUPPLANTS WC

The worthy successor to the venerable WC was the WD model, which first appeared in 1948. It offered more horsepower and

1946 Allis-Chalmers B
Lights and electric starting made the Model B a modern machine. This 1946 version has the later 125-cubic-inch engine that ran at 1,500 rpm for 19.51 drawbar horsepower and 22.26 belt horsepower. The rounded radiator, hood, fuel tank, and clam shell fenders were designed by Milwaukee industrial designer Brooks Stevens. He provided the Allis-Chalmers tractor look for many years. The Model B was made from 1937 until 1957, with a total of 120,783 rolling off company lines. That number includes some 2,800 of its low-slung IB industrial version. This 1946 Allis-Chalmers B is owned by Tom Humbert of McConnellsburg, Pennsylvania.

many modern features. The gasoline WD pulled 30.23 drawbar horsepower and 34.63 belt horsepower when tested at Nebraska in 1948, some 32 percent better than the tractor-fuel (distillate) burner.

The WD offered continuous power take-off by way of a two-clutch control. The transmission hand clutch stopped forward tractor movement while the PTO and hydraulic pump continued to run. People who owned the popular A-C Roto-Baler loved the feature. They could stop forward movement quickly while the baler finished forming, wrapping, and ejecting its bale. The hand clutch ran in oil and was forgiving of a lot of slipping when it was used as a speed control in tough harvest conditions.

Optional WD features included power-adjusted rear-wheel tread spacing, achieved by using spiral bars on the wheel rims. Operators could spin out the tires to ten different preset tread widths. The Power-Shift rear wheels were an A-C innovation and an industry first. Power-Shift-type tread adjustment was soon in

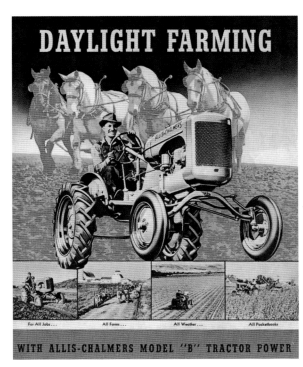

1940s Allis-Chalmers Model B brochure
In advertising its Model B, Allis-Chalmers said the tractor was "for all jobs . . . all farms . . . all weather . . . all pocketbooks." This brochure also shows the tractor's main purpose: to replace the last team of horses on the farm.

1939 Allis-Chalmers IB

Industrial users were the target market for the compact low-profile industrial Model IB, derivative of the Allis-Chalmers Model B. The rear drop gearboxes were turned 90 degrees forward to lower the axles and shorten the IB's wheelbase. A straight front axle lowered the front end. This 1939 IB shows the exterior channel iron frame that was used to mount equipment. Charlie English Jr. of Evansville, Indiana, collected this 1939 Model IB.

Allis-Chalmers G
Truck farmers, nurseries, and specialty crop growers were the intended market for this radical "wheeled hoe," introduced in 1948. The Allis-Chalmers Model G tractor put the operator right above the row it cultivated. Many different implements could be mounted on the two arched tubular frame elements near the front of the tractor. The Model G was powered by a small four-cylinder 62-cubic-inch Continental engine, producing 9.04 horsepower worth of drawbar push. This Model G is owned by Jim Hanna of Rochester, New York.

use industry wide when A-C licensed its patents. To no one's chagrin, the WD's brake pedals replaced the lever-operated brakes used on the WC.

The WD also offered a single hitch-point implement attachment under the rear axle. Connecting the hitch under the axle was a chore, but was soon remedied on the 1945 Model WD45.

Until 1952, a four-speed sliding-gear transmission was used on the WD. That year a new constant-mesh transmission replaced the older unit and made gear shifting easier. In 1953, magnetos for ignition were replaced with battery-powered distributors. The WD was available with a tricycle front, adjustable front axle, or fixed wide front axle. Dual, or tricycle-front, WD tractors listed for $1,685 in 1947 and rose in price to $2,060 in 1951, as post-war inflation took hold.

Allis counted the WD as another success for its W Series tractors. From 1948 through 1953, it sold 146,125 WDs. Yet there was still another, even better machine waiting in the wings.

WD45 POWERS UP

By the early 1950s, competing farm tractor makers were starting to offer row-crop tractors that approached 45 belt horsepower, or four-plow-rated tractors. The International Harvester Model M

1950 Allis-Chalmers G
Visibility for crop cultivation didn't get any better than on this 1950 Model G tractor. Careful cultivation of vegetables and other delicate crops could be achieved by slowing the tractor to .75 miles per hour in first gear. Here, Dwight Hughes Sr. cultivates nursery stock at the Dwight Hughes Nursery near Cedar Rapids, Iowa. Nearly 30,000 Model G tractors were built in Gadsden, Alabama, by Allis-Chalmers from 1948 to 1955.

Farmall and the John Deere 70 had upped the power ante, and Allis-Chalmers wanted its own 45-belt-horsepower machine.

The new A-C Power-Crater gas engine used in the new WD45 model boosted its Nebraska test ratings to 43.21 belt horsepower and 37.84 drawbar horsepower when it tested there in 1953. Those figures were about 25 percent more power than those from the old WD. That boost also gave the WD45 enough extra "muscle" to add another plow and consider it a four-plow machine.

Displacement on the WD45 engine was 226 cubic inches, 25 cubic inches more than the old WD engine had. Allis added a half inch to the engine's crankshaft throws to get the extra displacement. The four-cylinder engine's bore stayed at four inches and its

1952 Allis-Chalmers WD
After World War II, Allis-Chalmers answered the demand for more modern and more powerful tractors with its Model WD, the replacement for the venerable Model WC. Introduced in 1948, the WD offered continuous PTO by way of a transmission hand clutch. PTO-driven implements like this mounted No. 33 corn picker could catch up or grind out a plug up while ground transport was stopped or slowed. The WD boasted 30.23 drawbar horsepower, a spin-out or Power-Shift rear-wheel tread width setting, and even brake pedals. Duane Eilts of Massena, Iowa, put together this 1952 WD and mounted corn picker.

rpm was kept at 1,400, but the new engine's compression ratio was raised to 6.45:1. A newly designed combustion chamber shape prompted the Power-Crater designation.

Help with hitching integral implements came when Allis-Chalmers developed the Snap-Coupler for the 1954 Model WD45. A sturdy-cast funnel guided the notched implement tongue into the spring-loaded latch. All the operator had to do was back the

tractor carefully into the hitch, wait for its "snap," attach the hydraulic lift arms, and then drive off with the implement.

In 1956, the WD45 became the first Allis-Chalmers tractor to have factory power steering. Was there anything it still lacked? A diesel engine, of course. Cost-conscious farmers were increasingly looking to diesel-powered tractors to reduce fuel costs. A-C's competitors were offering diesels, so the manufacturer wanted to offer one, too.

WD45 DIESEL

The WD45 became the first Allis diesel wheel tractor in 1955 after some quick action. In 1953, Allis Chalmers acquired the Buda Company of Harvey, Illinois, a diesel engine manufacturer, and fit its six-cylinder 230-cubic-inch diesel into the WD45. By October 1954, the WD45 diesel tractor went into production and 350 of them were made before the year ended. Nebraska tests on the diesel WD45 in 1953 showed it could produce 43.29 horsepower from the power take-off.

By the end of the WD45's production life in mid-1957, some 90,382 copies of the tractor had been made, including 84,030 WD45s and 6,489 WD45 diesels. After 1953, Allis quit making distillate-fueled engines.

In a marked switch, the adjustable front-axle version on the WD45 outsold the tricycle-type front, the previous farmer favorite for wheel placement on row-crop machines.

CA TWO-ROW MODEL

A somewhat more powerful version of the light two-plow Model C, the CA, was introduced in 1949. Its engine power was increased by boosting its rpm to 1,650 and increasing the compression ratio to 6.25:1.

The CA was a two-row, two-plow tractor that was available with a tricycle front or adjustable wide axle, and it offered other improvements over the Model C. Its four-speed constant-mesh transmission allowed easier shifting into its speeds of 2, 3.5, 4.5, and 11.25 miles per hour. Nebraska tests pegged the Model CA at offering 25.96 belt horsepower and 22.97 drawbar horsepower from its gasoline-fueled engine, a substantial improvement over the Model C.

The CA had the features first seen in the larger WD model. These included Power-Shift wheel-tread adjustments on its 24-inch rear wheels and the Allis-Chalmers Snap-Coupler. The CA's ignition now utilized a distributor system, a change that also enabled the tractor to be equipped with electric lights and a starter. From 1950 to 1958, production of the CA reached about 39,500 tractors.

THE D10 AND D12

The B and CA models were replaced in 1959 when production started on two new models: the D10 and D12. Only the wheel treads were different between the two-plow models. The D10 was a short-axled one-row machine, while the longer-axled D12 could spread its tread to handle two-row chores. The models shared a common engine and drivetrain.

The engine had a 138.7-cubic-inch displacement with a 3.375x 3.875-inch bore and stroke. It was tested at Nebraska in the D10 in 1959 and achieved 28.51 PTO horsepower and 25.84 drawbar horsepower at 1,650 rpm while fueled with gasoline. The D12 tests came up with basically the same figures.

The D10 and D12 were painted in the old Persian Orange color with black trim accenting their radiator grilles until late in 1960, when they appeared in a new shade of Persian Orange paint with cream wheels.

The D10 and D12 added Series II to their designations in 1963 after engine, PTO, and hydraulic improvements. Their horsepower got a boost, and their PTO systems became "live" and were finally matched to industry standards. A new gear-driven live pump that standardized its lower 1,500 psi pressure with other lines of equipment drove the hydraulics. Engine changes added pull to the models when the displacement on the four-cylinder engine went to up to 149 cubic inches.

Further refinements resulted in the D10 and D12 being called Series III machines beginning in 1964. Most important of those changes was the availability of a two-range transmission. It doubled the operating speeds of the two tractor models to eight. The D10 and D12 models were produced from 1959 to 1968, in which time 5,304 D10 units rolled off factory lines and 4,070 D12s were produced.

HI HO HORSEPOWER—THE RACE IS ON

Horsepower, the more the merrier, was the big game in the farm tractor business in the late 1950s. America's farmers were competing with each other for land and for top crop yields. So operators needed more powerful, more efficient tractors to grow more bushels. The U.S. tractor makers were challenging each other to provide the much-needed power and nudge ahead in the race. Power "tweaks" were rampant.

Allis-Chalmers responded to the challenge with a complete redesign of its tractors, first introducing the Model D14. It came out in early 1957. The larger Model D17 soon followed.

1955 Allis-Chalmers WD45 Diesel
On its way to 45 horsepower and a four-plow capacity, the Model WD45 made many farmer friends after its 1953 introduction. Buyers with fuel economy in mind were attracted to this 1955 diesel version. Its six-cylinder diesel engine was an Allis-Chalmers first. The 230-cubic-inch diesel gave the WD45 tractor 39.5 drawbar horsepower. Built only between 1953 and 1957, more than 90,000 WD45s sold. Diesels made up 6,489 of the total. Adding power steering to the 1955 model—this one collected by Kenny Kamper of Freeburg, Illinois—added even more farmer appeal.

1950s Allis-Chalmers D Series brochure
Advertising the D14 and D17, this A-C brochure promotes its D tractor series' power as a way toward better living, better farming, and more profit.

1958 Allis-Chalmers D-17
This 1958 Model D-17, restored by Kenny Kamper of Freeburg, Illinois, has its operator's station straddling the transmission and in front of the differential housing, a D Series feature. A three-position Power-Director lever gave the D-17 two speeds in each of the four transmission gears. Full forward, the lever provided the high-range speed, while full rearward, the travel speed was reduced. In the central neutral position, the tractor stopped but kept the PTO and the hydraulic pump running. Careful use of the Power-Director allowed on-the-go gear shifting. The D-17 had a 226-cubic-inch gas engine that produced 52.7 belt horsepower.

1963 Allis-Chalmers D-12 Series II
This 1963 A-C Model D-12 Series II has independent PTO and live hydraulics. Its four-cylinder 149-cubic-inch gas engine made it a strong two-plow tractor with 29.43 horsepower available at the drawbar. The three-point hydraulic hitch was draft-sensitive. Its four-speed constant-mesh transmission gave it travel speeds of 2, 3.5, 4.5, and 11.4 miles per hour. This tractor, owned by Howard DePrenger of Otley, Iowa, is painted in Persian Orange No. 2, and the wheels and radiator grille are cream color.

Using a trend first set by the grey Ford-Ferguson of 1939, the D14 had the operator seated slightly in front of and above the new tractor's differential case, with his legs straddling the transmission housing and his feet just inches away from the wheel brake pedals on the right and the clutch pedal on the left. Within easy reach of the operator's right hand was a new control: the Power-Director hand clutch. Its center position stopped the tractor with the hydraulics and the PTO still running. That neutral position also allowed for shifting gears on the fly. In its forward position, the Power-Director gave the tractors high-range speed. In the rear position, the speed was reduced by 30 percent to help pull out of tough spots. That gave both new tractor models eight forward speeds and two reverse speeds.

1967 Allis-Chalmers D-10 Series III
Taking on one crop row at a time was the aim of the short-axled D-10 Series III model. It shared the same engine and powertrain of the two-row D-12 and put out 28 drawbar horsepower and 33 PTO horsepower. This 1967 D-10 is owned by Howard DePrenger of Otley, Iowa.

The really big news in launching the D14, though, was its new 3.5x3.87-inch four-cylinder engine. The D14 used the Power-Crater combustion chamber on its 149-cubic-inch gasoline engine to earn 34.08 belt horsepower and 30.91 drawbar horsepower at 1,650 rpm at its Nebraska tests. Those numbers were far better than those on the WD45 model it replaced. The D14 was a strong two- to three-plow tractor.

During its production run from 1957 to 1960, Allis manufactured 22,292 D14s. The model was replaced with the D15 in 1960. The D15 production amounted to 17,474 tractors between 1960 and 1968.

Power-steering hydraulics helped adjust front-tread settings on the adjustable front ends of the D14 and D17 models. Allis called the feature the Roll-Shift adjustable front axle. The models were both available with dual-, single-, or adjustable-front axles on their front ends

The new Model D17, which came out in 1957, also received a power infusion to its 226-cubic-inch, four-cylinder Power-Crater engine by increasing the compression ratio to 7.25:1 and boosting the rpm to 1,650. The gas-version D17 produced 52.70 belt horsepower at Nebraska, while the diesel D17 put out 51.14 belt horsepower and 46.20 drawbar horsepower in its 1957 Nebraska tests. The new D17 six-cylinder diesel had also been tweaked up to 262 cubic inches of displacement and its rotation rate was raised to 1,650 rpm. The D17 replaced the WD45 in the A-C lineup and gave Allis a two-wheel-drive tractor of more than 50 PTO horsepower.

Total production of the D17 reached 51,598 for the gas machines and 12,284 for diesel D17s, which were made from 1957 to 1967. When the model was retired, there was still life left in the D17 design. It would become stronger *and* taller as the Model D19.

PERSIAN ORANGE NO. 2

Persian Orange No. 2 wasn't the only paint feature that made the Model D15 distinctive in the fall of 1960. The D15s rolling off company lines also had cream-colored trim. The new orange paint being used was a "redder" tint, and the tractor was stronger, too.

The D15's extra horsepower came from a compression ratio increase to 7.75:1 and its engine being speeded up to 2,000 rpm. The gasoline-version D15 that tested in Nebraska that year turned out 40.00 PTO horsepower and 35.94 drawbar horsepower. Its diesel sibling produced 36.51 PTO horsepower from its 175-cubic-inch engine at 2,000 rpm. The diesel engine was a four-cylinder

design that A-C based on the six-cylinder diesel used in its D17. The D15's diesel engine had a 3.562x4.375-inch bore and stroke and a 15.5:1 compression ratio.

In a concession to the times, Allis offered a three-point hitch as well as its own Snap-Coupler on the D15. The model became the D15 Series II, introduced early in 1963. Its gas engine was increased to 160 cubic inches of displacement by a 0.125-inch increase in cylinder bore. The D15 Series II had tested 46.18 horsepower on the PTO and 38.33 drawbar horsepower. The diesel D15 Series II kept its old 36.51-PTO horsepower engine.

D19 WITH BLOWER TO GO

More horsepower for the D17 was at the top of Allis-Chalmers' wish list in 1959. John Deere had just stunned the industry with its advanced New Generation series of tractors. The Deere Model 3010 sported 54 drawbar horsepower; the bigger six-cylinder 4010 bragged of nearly 74. The Allis D17 still only produced 46 horsepower at the drawbar, well below the Deere 3010.

Weighing its options, Allis-Chalmers forged ahead with a bold plan and created the 1961 Allis-Chalmers Model D19 diesel. It showed 66.96 PTO horsepower and 61.27 drawbar horsepower in 1961 Nebraska tests, well ahead of the Deere competition. The gas version D19 proved even better, cranking out 71.54 PTO horsepower.

What magic had spurred such success? A turbocharger. Its introduction on the D19 was the first use of a factory-installed turbocharger by a U.S. tractor maker. The D19 diesel also was the first turbocharged tractor tested at Nebraska.

Turbochargers, using engine exhaust gases as a power source, compress engine combustion air, allowing engines to produce more power. At the time, farmers could add turbochargers to their tractors as add-ons from outside suppliers, but the use of them usually voided tractor manufacturer's powertrain warranties.

The D-19 was the biggest tractor in the A-C line. It stood tall on its new rear 34- or 38-inch rims and bigger tires. The offset rear gears and small rear tires that had been in use since the introduction of the WC model were now gone. New single-piece rear axles carried power to the wheels. The tractor hood line was raised to allow a larger fuel tank. Power steering came standard on the 7,800-pound tractor. A Power-Director hand clutch, similar to that on the old D17, gave its driver the old Allis feeling.

1966 Allis-Chalmers D-15 Series II
This 1966 D-15 Series II produced 38.33 drawbar horsepower and 46.18 PTO horsepower in Nebraska tests, running its 160-cubic-inch engine at 2,000 rpm. The Power-Shift control gave the D-15 eight forward speeds. An optional three-point hitch made it an attractive alternative to buyers who had other makes of equipment to mount. Four-row equipment was easily managed by the D-15 Series II. This 1966 D-15 was collected by Kenny Kamper of Freeburg, Illinois.

Allis also made a high-crop version of the D19, which provided 37 inches under-axle crop clearance. The D19 was manufactured from December 1961 to mid-1964 with a run of 6,955 gasoline tractors and 3,625 diesels.

THE D21'S 100 GALLOPING HORSES

While not the first across the 100-horsepower line, the new 1963 D21 Diesel was a close contender. The D21 tested in 1963 at Nebraska as a 100-plus-horsepower brute. The figures show it produced 103.06 PTO horsepower and 94.05 drawbar horsepower. Its power came from a new six-cylinder, 426-cubic-inch, Model 3400 direct-injection diesel engine. The engine rolled ahead at 2,200 rpm and had a 4.25x5-inch bore and stroke. The D21 Diesel had four forward gears in its new constant-mesh transmission and two-speed range, from to 1.6 to 16.2 mile-per-hour gaits. The new tractor had a speed for every need!

Allis-Chalmer's first big new tractor of the era was styled by the company's new industrial design department. The designers put a big operator's platform above the gear case, allowing the operator to use the mighty machine in a standing position and take a break from long hours in the tractor seat. Hydrostatic steering, with a tilt wheel and instrument panel attached, kept the gauges in view as the steering wheel was raised. Large fenders, with lights mounted top front, kept mud and loose dirt off of the operator. With a big 52-gallon diesel tank, the D21 could work for nearly ten hours without refueling.

A turbocharged diesel Model 3500 Allis engine, introduced in 1965, turned the D21 into an even stronger stallion. The horsepower thus jumped to 127.15 on the PTO and 117.57 on the drawbar in the Nebraska tests. That was a 24 percent increase in power using only 7 percent more fuel. Known as the D21 Series II, the tractor was tested at a weight of 10,675 pounds.

The D21 Series II was made in the West Allis plant, where 2,329 rolled off the line from 1965 until 1969. The first D21 model was also made at West Allis, from 1963 until 1965, with 1,128 made. The two D21 Series models combined totaled 3,457 tractors.

TO THE END

Allis-Chalmers made its line of Hundred Series tractors from 1964 until 1981, its 7000 Series from 1973 to 1981, and the 6000 and 8000 four-wheel-drive tractors from 1980 to 1985. And that's when it all headed downhill.

Allis had progressed from a beginner in the farm equipment business with its first farm tractor in 1914 to the third-largest tractor maker in 1936 sales. It continued as a worthy competitor of the other major manufacturers with innovative designs and creative marketing through the next forty years. But the farm crisis of the 1980s hurt the company, and in December 1985, after a run of more than seventy years, tractor production stopped at West Allis, Wisconsin.

Deutz-Allis, the successor to Allis-Chalmers, was formed under the ownership of Klockner-Humboldt-Deutz (KHD) of Cologne, Germany. Under Deutz auspices, the tractors sold in the United States were green air-cooled Deutz machines. In 1990, Allis-Gleaner Company (AGCO) bought out Deutz-Allis. In the intervening years, AGCO, headquartered in Duluth, Georgia, has continued to buy and merge farm equipment companies worldwide. In 2005, it reached $5.5 billion in sales as the world's third-largest agricultural equipment supplier. It claims a ten percent share of the North American market.

AGCO products are distributed in more than 140 countries under well-known brand names, including AGCO, Challenger, Fendt, Gleaner, Hesston, Massey-Ferguson, New Idea, and others. Those painted in familiar Persian Orange, the AGCO-Allis line, could once be found in AGCO's stable. Now solely AGCO machines, these tractors come in red. AGCO's Challenger division, once Caterpillar's agricultural tractor line of rubber-tracked crawlers, will soon have a new line of 430- to 550-horsepower articulated rubber-tired four-wheel-drive tractors. They will be powered by Caterpillar engines.

Opposite: **1968 Allis-Chalmers D-21 Series II**
A turbocharger first breathed more air and more power into the Model D-21 Series II diesel in 1965. This 1968 model, collected by Kenny Kamper of Freeburg, Illinois, spun its 426-cubic-inch six-cylinder diesel engine at 2,200 rpm to produce 117.57 drawbar horsepower. With that amount of power, the D-21 could pull seven to eight plows through the fields. A heavy-duty four-speed constant-mesh transmission, with the help of its Power-Director, gave it eight forward speeds and two in reverse. Hydrostatic steering, a comfortable operator's seat and platform, and a big 52-gallon fuel tank helped the D-21 put in long, productive days.

Chapter Three
J. I. Case

1939 Case R
This 1939 Model R Case is a standard-tread version. A Flambeau Red color, new streamlined sunburst grille, and rounded clamshell fenders marked the Model R, the RC, the orchard RO, and the industrial RI beginning in 1939. Electric lights and starter were available on the Model R in 1939. This tractor, on steel wheels, was exported to Canada. Rubber tires were also available. Robert J. Porth of Regina, Saskatchewan, counts this Model R as part of his tractor collection.

Jerome Increase Case founded his company in 1844 and quickly grew it into the world leader in threshing machines and steam engines before his death in 1891. The J. I. Case Threshing Machine Company brought the threshing machine to the Midwest and built about a third of the North American steam traction engines made between 1876 and 1925.

Moving from his home in western New York in 1842, Case came to the heart of the new Midwest grain-growing country as a young man of twenty-two. He brought with him six "ground hog" threshers, crude machines with revolving pegged wooden cylinders that threshed the grain from wheat as it was hand fed into the thresher. The wooden contraptions were a vast improvement from hand flailing grain on a threshing floor. Case's father, Caleb, sold the threshing devices and Jerome, Caleb's youngest son, apparently bought the machines on credit, most likely from his father.

On his way to Rochester, Wisconsin, from his home at Williamstown, near Rochester, New York, the young Case sold five of his "ground hog" machines in the fall of 1842. The sixth machine he kept for himself, first to use for custom threshing and then to remake into a machine with more capabilities. It became the foundation of his business as he coupled it with a "fanning mill" and created his first crude threshing machine. It was ten times faster at threshing than the old method of barn floor threshing.

RACINE MACHINES

By spring of 1844, J. I. Case had successfully demonstrated his unit, and by the end of the year, he had moved to Racine, Wisconsin, opened a shop, and had six threshing machines built. Employing patent rights from other threshing machine makers of the day along with his own ideas, Case soon led the field, and his machines won coveted prizes at the agricultural expositions of the era.

On completion of his new three-story 30x80-foot brick shop in 1847, Case wondered whether he would ever fill it with thresher production. That was not a problem, though. The Civil War era in the early 1860s was a real boon to early farm equipment makers like Case. Farm mechanization kicked into high gear because farm help was scarce as young men took up arms.

J. I. Case took on three partners in 1863 and operated as J. I.

1900s Case Steam Tractor advertisement
Jerome Increase Case grew his company into the world leader in steam engines beginning in 1844 up to his death in 1891. The first Case steam engine arrived in 1876.

1918 Case 9/18A
Case really got its gas tractors rolling with this 1918 Model 9/18A. It set the pattern for the crossmotor line of Case tractors that followed. The small tractor was aimed at a market that had been dominated by much heavier machines. The 9/18 featured a transverse-mounted four-cylinder engine of 3.875x5-inch bore and stroke. Its two-gear transmission propelled it at 2.25 and 3.5 miles per hour. Its weight of 2,650 pounds was considered to be light for the era. Intricate pin striping on its green with red trim paint job gave it a finished look. This early Case gas tractor is owned by Herb Wessel of Hampstead, Maryland.

1919 Case 10/18
Belt power flowed from the belt pulley on the end of the crankshaft in the 1919 Case Model 10/18. It had the same engine as the 9/18, but with an increase in engine rpm, the 10/18 produced more horsepower. It was a two-plow tractor that could pull three under ideal conditions. Case called it the "year-round tractor" and stopped making the three-wheel 10/20 crossmotor as this model entered the market. This 1919 Case 10/18 is owned by John W. Davis of Maplewood, Ohio.

Case and Company. The partnership allowed him to expand the Case business to meet the new demands. The partners were Massena Erskine, Robert Baker, and Stephen Bull. Along with Case, they became known as the Big Four.

In 1880, the Big Four partnership of J. I. Case and Company was dissolved, and the business was incorporated as the J. I. Case Threshing Machine Company. In the ensuing years, Case became "king" of threshing machines, and later his company became "king" of farm steam engines. Case made and sold 35,737 steam traction engines between 1878 and 1924. That production was more than one-third the total 83,824 such engines made in the United States. The second-largest producer, Huber of Marion, Ohio, made some 11,000 engines.

HORSE-POWERS

Power for the threshing machines Case built was initially supplied by single-horse treadmills. Later came "horse-powers," which harnessed teams of horses behind "sweeps" of revolving levers geared to a "tumbling" shaft. The shaft transferred the horse power to the ever-improving threshing machine. Case made horse-powers matched to the power requirements of his threshing machines. One J. I. Case–made Dingee-brand horse-power harnessed seven teams, or fourteen horses, to one J. I. Case thresher.

RACINE STEAM MACHINES

The first Case steam engine, now enshrined at the Smithsonian Institute in Washington, D.C., came out in 1876. It was a horse-drawn engine of 10 horsepower, designed to give the horses a break in powering the threshing machines of the day. The Case portable steam engines became self-propelled shortly thereafter, but weren't hand steerable until 1884. Until then, a team of horses was still used to guide them.

J. I. Case died in 1891 and was succeeded as president of the J. I. Case Threshing Machine Company by Stephen Bull, his brother-in-law.

Case steam engine production peaked in 1912 as gasoline-powered tractors started to roll out of Case factories. By 1924, the

1920s Case 10/18 advertisement

This Case advertisement for the 10/18 highlights the features of the tractor's chassis in a detailed diagram so that farmers could see how many of the 10/18's parts were easily accessible if they needed to be fixed.

market for steam power had fizzled, and Case steam engine production stopped after a total production run of more than 35,000 farm engines.

CASE TRIES GASOLINE ENGINES

The Case company's first pioneering efforts at producing a gasoline traction engine date to 1894, or maybe earlier, when William Paterson of Stockton, California, came to Racine to make an experimental engine for Case. Case advertisements in the 1940s, harking back to the firm's history in the gas tractor field, claimed 1892 as the year Paterson designed his gas traction engine. Patent dates suggest 1894. The early machine ran, but apparently not well enough to be produced. Carburetion and ignition devices lagged behind the other mechanisms used and thus limited its dependability.

Paterson's "balanced" engine design was novel, if not entirely successful. It was a horizontal two-cylinder inline engine with the water-jacketed combustion chambers adjoining head to head (but not connected) in the engine center. Its crankshaft was connected directly to one piston by a connecting rod and indirectly connected to the opposite piston through a walking beam and two parallel-acting connecting rods. William Paterson and his brother James shared the original engine patent dated October 30, 1894.

After deciding not to produce Paterson's design, Case stayed out of the gas tractor business until 1911 when it returned with its 30/60 model.

The big 30/60 model won Case first place at the Winnipeg tractor trials in 1911. It was a monster of nearly 13 tons and was produced from 1912 through 1916. A smaller version, a 12/25 horsepower tractor, came out in 1913. A mid-sized 20/40 won Case more gold medals at the 1913 Winnipeg trials. Like many tractors of the era, these three models were powered with two-cylinder opposed horizontal engines. Crankshafts were offset 360 degrees to produce a power stroke every revolution. Many components of the heavy machines were from Case's steam line, including rear drive wheels, gearing, and hot-riveted frames and chassis.

In less than a decade, Case moved from making massive steam-era two-cylinder engines to lighter automotive-type tractors with four-cylinder vertical (but transverse) engines. Case tenaciously held to its view of the superiority of spur gears over bevel gears for power transmissions until 1929, when the Model L gas

Emerson-Brantingham advertisement
This colorful Emerson-Brantingham advertisement aims to attract both farmers and agricultural equipment dealers by noting E-B tractors and implements are "keys to successful farming and sure profits to the dealer."

1916 Emerson-Brantingham L
This 1916 Emerson-Brantingham Model L was not a Case tractor, but a machine made by a company Case bought in 1928. The L was discontinued by the time Case bought E-B and was one of the few tractors farm equipment pioneer E-B designed. This three-wheeler was rated at 12/20 and was said to be able to pull a three-bottom plow. It had forward speeds of 1.66 and 2.33 miles per hour. The operator sat and steered on the furrow side of the E-B's big drive wheel. The Model L was built for only a year. This Model L is owned by Dennis and Susan Black of Arlee, Montana.

Case L

The Case crossmotor line ended in 1929 when it was superseded by the modern Model L. It was a new design with a four-cylinder engine arranged inline with the tractor frame. The new long-stroke valve-in-head engine had a 4.624x6-inch bore and stroke. It proved 47.04 belt horsepower during Nebraska tests. Painted gray instead of the previous Case dark green, the 26/40 Model L standard-tread machine set the pattern for later tractor designs like the Model C and the row-crop 17/27 Model CC of late 1929. David T. Erb of Vinton, Ohio, owns this Case L.

1930s Case Model L brochure

tractor was introduced. Cross-mounted (or transverse) engines made bevel gears unnecessary, since the engine, drivetrain, and wheels all rotated in one plane.

THREE WHEELS

The first Case crossmotor engine design followed two trends of its time: that of the three-wheeled tractor and of the lightweight machine. Like some other popular makes of the day, the Model 10/20, introduced in 1915, had one large driving "bull" wheel on the right or furrow side with its front steering wheel aligned with it on the right. The idler wheel on the left, or land, side had no differential but could be temporarily clutched into the live axle for extra traction in tough going.

The 10/20 was a two-plow tractor weighing just over 5,000 pounds, a true lightweight in 1915. Its second-generation engine,

an overhead-valve four, was the basis for later crossmotor designs. Production of the 10/20 continued until 1918, and it was still being sold as late as 1924. About 5,000 three-wheel 10/20s were sold in the United States before the larger 12/20 crossmotor replaced it in 1921. The 12/20 model sold about 11,500 machines from 1921 to 1928.

THE SMALLER THE BETTER?

With Henry Ford's Fordson threatening to gobble up much of the market for lightweight tractors, Case quickly brought out another lightweight tractor, its 9/18 crossmotor, in early 1916. It weighed in at about 3,650 pounds—a little more than a team of horses. It was rated as a two-plow machine and was advertised as being able to handle a small thresher.

A few more than 6,000 of two versions of the 9/18—models 9/18A and 9/18B3—were made before its production run ended in 1919. It was not a big seller when compared with the Fordson, but it foreshadowed a complete line of Case crossmotor tractors, which eventually grew in size back up to the horsepower ratings of Case early gas-oil machines.

By 1917, the 10/18 model was on the market. Similar to the 9/18B, but with a cast radiator tank and higher engine rpm, the 10/18 passed a production run of 9,000 units before its manufacture ended in 1920. The 10/18 was the third tractor tested at the new Nebraska test station in Lincoln, Nebraska. Records of the tests show the 10/18 produced 18.41 belt horsepower and 11.24 drawbar horsepower at 1,050 rpm in April 1920.

Case Model 15/27 came out in 1919 and soon was the company's best seller. Rated at three-plow capacity, it hit a responsive chord with buyers. The 15/27 was the first Case tractor equipped with a power take-off (PTO) to power implements hitched to its drawbar. Nebraska test No. 4 in 1920 placed the 15/27 well within Case's rated horsepower. Tested at 6,460 pounds, the tractor proved 27.52 horsepower on the belt and 18.80 horsepower at the drawbar. Its 4.5x6-inch bore-and-stroke engine turned at a leisurely 900 rpm. The 15/27 sold more than 17,600 units before it was replaced with the upgraded 18/32 model in 1924.

Also new to the market in 1919 was the Case Model 22/40. The 22/40, which could pull four or five plows, weighed more than 9,000 pounds. Measured horsepower of 40.80 on the belt and 23.51 on the drawbar were shown in its 1920 Nebraska tests. Its big transverse 5.5x6.725-inch bore-and-stroke engine turned at

850 rpm. By its production end in 1924, some 1,850 Model 22/40s were made. It was upgraded to 25/45 horsepower in 1925 with a few engine tweaks.

In 1922, Case replaced its first "lightweight" three-wheel 10/20 with the Case 12/20. Its pressed steel wheels were a ready identifier for the 4,230-pound cast-frame tractor with valve-in-head engine. The Model 12/20 had a 4.125x5-inch engine that ran at 1,050 rpm. It tested in 1923 Nebraska trials at 20.16 belt horsepower and 12.83 drawbar horsepower.

The Model 12/20 stayed in production until 1928 when it was redesignated as the Case Model A. Records show 9,237 Model 12/20s were made during its lifetime, making it second-most popular of the crossmotors. Unlike the green tractors with red wheels of the earlier crossmotors, the 12/20 model was painted all gray.

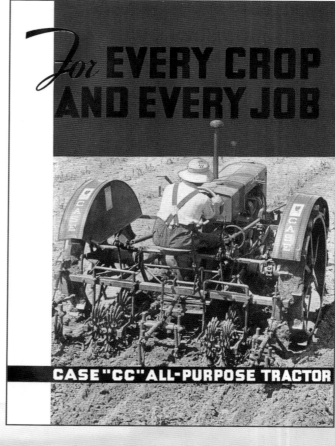

1930s Case CC All-purpose Tractor brochure

1935 Case CC
The Case Model CC was the Racine, Wisconsin, company's entry into the row-crop tractor market. It was based on the standard-tread Model C. Both tractors were considered 17/27 horsepower machines. Two- and four-row cultivators with motor lift were available for the tractor. Case kept the "chicken perch" side steering arm on its row-crop machines into the 1950s. This 1935 tractor, found and restored by Stanley Britton of Athens, Alabama, is equipped with its original steel wheels.

1936 Case CC

Rubber tires on this 1936 Model CC, owned by A. K. Gyger of Lebanon, Indiana, practically converted it into a new tractor. On rubber tires, the CC could pull a three-bottom plow in most conditions and was a much easier machine to operate. The longtime J. I. Case trademark—Old Abe, the eagle perched on the globe— is prominent on the tractor gas tank. Rear tire tread on the CC was adjustable from 48 inches (for plowing) to 84 inches (for row-crop work).

The largest Case gas tractor of the early 1920s was the Model 40/72. But during its three-year production run, only forty-two were made. Tested at Nebraska in 1923, the 11-ton machine produced more than 91 brake horsepower, a record for that time.

But it set another record, too. Its thirsty kerosene-burning engine gulped down more than 7.5 gallons per hour for a record low figure of 5.61 horsepower hours per gallon of fuel burned. It proved to be too much, too late. Smaller, more efficient designs were coming.

THE END OF THE CROSSMOTOR ERA

When Leon R. Clausen joined Case as president in 1924, the company was still firmly in its crossmotor stage. Clausen, previously vice president of manufacturing at Deere & Company, had pushed successfully at Deere for the development and manufacture of John Deere's popular two-cylinder Model D (introduced in 1923) and had started to help Deere pursue its row-crop or general-purpose tractor.

IHC's Farmall row-crop tractor had just been introduced and was catching the interest of a growing number of farmers and tractor manufacturers. Clausen wanted to move Case forward, toward more modern tractors, including row-crop types like the innovative Farmall. Clausen wanted a tractor design that could be used in a row-crop tractor because Case's crossmotors were too wide.

By the early 1920s, most tractor firms had already changed to automotive-type four-cylinder engines, which were aligned lengthwise along the frame. Case's transverse engine (crossmotor) mounting with its wide stance clearly wasn't adaptable to row-crop design, since the engine needed to fit in the 42 inches between the crop rows. Clausen stopped all crossmotor development at Case soon after he arrived and directed the firm toward newer machine designs. Those efforts ultimately resulted in the Model L and Model C tractors, as well as Case's late entry into producing row-crop machines.

ACQUISITION AFTER ACQUISITION

J. I. Case Threshing Machine Company became just the J. I. Case Company, Inc. in 1928, after Massey-Harris purchased the J. I. Case Plow Works. The plow company had become distinctly separate from the threshing machine company, and the two Case firms were often at odds with one another.

By purchasing the J. I. Case Plow Works, Massey-Harris acquired the Wallis tractor line, developed originally by Henry M. Wallis,

1936 Case RC
Designed to compete with the Farmall F-12, the Case Model RC was a two-row 18/20 tricycle. Case painted the RC light gray so customers wouldn't confuse it with the company's more powerful (and more profitable) Model CC. The early RC's over-the-top steering also was different from the side-arm arrangement on the CC. The RC's Waukesha four-cylinder engine produced 14.21 drawbar horsepower and 19.8 belt horsepower in Nebraska tests. John Bourque of St. Genevieve, Missouri, restored this 1936 Model RC that replaced mules on his home farm in the Mississippi bottoms.

1937 Case CCS

This Model CCS was a special row-crop designed to cultivate high-growing crops. This 1937 tractor was shipped from the factory in October 1937 and worked in Louisiana's sugar cane fields. The bowed front axle raises the front of the tractor, and drop gearboxes add crop clearance under the rear axle. The special deep lugs on the rear wheels weigh about 21 pounds each. Charles Q. English Sr. of Evansville, Indiana, restored this high-crop Case, which has the longtime trademark of the J. I. Case Threshing Machine Company cast in the differential housing.

Above: 1911 Heider A

The heritage of the early Heider tractor became part of Case history in 1937 when Case bought the distressed Rock Island Plow Company of Rock Island, Illinois. The Heider tractor line, started in Iowa in 1908 by Henry J. Heider, became part of the Rock Island firm in 1916. This 1911 Heider Model A, built in Carroll, Iowa, used a large friction plate to transfer engine power from the four-cylinder engine through a spur-gear transmission to the drive wheels. Only twenty-five of this model were made. Its Marion, Indiana–made Rutenber engine had a 4x4.75-inch design and ran at 850 rpm. Evaporative cooling dispersed engine heat. Dick Collison and Omar Langenfeld of Carroll, Iowa, restored this early machine.

Right: 1910s Rock Island Plow Company advertisement

This early tractor advertisement promoted the Heider Model C, 12/20 horsepower machine, and the Model D, 9/16 horsepower machine. It noted that Heider tractors could be trusted because they had experienced "nine years of actual field work."

1915 Heider C
By 1915, this Heider Model C friction-drive tractor was being produced in Carroll, Iowa. The tractor's variable speed friction-drive controls moved the engine forward or backward to adjust travel, or belt speed. The Model C's 4.5x6.75-inch Waukesha engine made three-plow power at a rated 10/15. A similar model, the D, was a two-plow tractor. Ground speed was varied by moving the engine forward or backward to change the effective radius of the friction plate where it contacted the engine flywheel. Dick Collison and Omar Langenfeld of Carroll, Iowa, restored this 1915 Heider C.

J. I. Case's son-in-law and president of the plow company. Massey then sold the rights to the J. I. Case Plow Works name back to the J. I. Case Threshing Machine Company and J. I. Case Company became the shortened name of the surviving operation.

The J. I. Case Plow Works had been formed back in the steam era to make plows for Case steam traction engines. After J. I. Case died in 1886, the plow company and the threshing machine company grew apart and friction between the two companies was palpable.

Acquisitions weren't a novelty at Case. One of its largest purchases came in June 1928, when it bought the huge, but financially troubled, Emerson-Brantingham Company of Rockford, Illinois. The addition of E-B gave Case a long line of well-known implements, a valuable dealer network, and new sales territory in the heart of the Corn Belt.

Emerson-Brantingham, like Case, was a pioneer in the agricultural equipment field and traced its roots back to 1852 and John H. Manny's reaper. In 1852, Manny beat out the acclaimed inventor of the reaper, Cyrus McCormick, in a field competition in Geneva, New York. Medals in hand, Manny continued to make and sell reapers and by 1854 had moved his operation to Rockford, where he took on partners. First to join were Walter and Sylvester Talcott in March 1854. By year's end, the trio had produced 1,030 Manny reapers. In September, the young company was joined by Blinn and Emerson, the Rockford hardware firm that was selling the Manny reaper. Hardware-man Ralph Emerson was a cousin of poet-essayist Ralph Waldo Emerson.

Manny died rather suddenly at the age of thirty, and company reorganization followed. Talcott, Emerson & Company became Emerson and Talcott in 1871. In August 1895, Charles S. Brantingham became secretary and general manager. Talcott died in 1900 and the company became the Emerson Manufacturing Company. Growth followed. In 1909, Charles Brantingham became company president, and the firm became the Emerson-Brantingham Company. Emerson was elected chairman of the board and rapid expansion followed.

1929 Rock Island FA
This 1929 Rock Island Model FA, owned by Richard E. Bockwoldt of Dixon, Iowa, was a derivative of the earlier Heider line. In the intervening years, Rock Island Plow Company changed the tractor's design from using the original friction drive to using a conventional clutch and gear transmission. The FA got its power from a four-cylinder Buda 4.5x6-inch engine that was rated at 18 drawbar horsepower and 35 belt horsepower at 1,100 rpm. Rock Island quit making tractors before it was acquired by Case in 1937.

Case Model D Series brochure

1939 Case DC

"Full three-plow power to hustle the heavy work and keep ahead of the weeds and weather" was promised by Case ads announcing the Flambeau Red Model DC in 1939. It was an update of the earlier Model C tractors. On steel, the DC price was $1,040; on rubber, it was $1,270. A rounded cast radiator grille and a rounded hood with a chrome strip accent streamline this 1939 model, collected by Warren Kemper of Wapello, Iowa. Lights and a starter are part of this tractor's installed options. Dual fuel filler caps suggest this machine burned distillate with help from gasoline at start up.

Starting in 1912, E-B added the Pontiac Buggy Company, the LaCrosse Hay Tool Company, the Rockford Gas Engine Works, the Geiser Manufacturing Company, Reeves and Company, the Gas Traction Company, the Newton Wagon Company, the American Drill Company, and others. All were moved to Rockford and into the vast three-story manufacturing complex E-B built there. Spread over 200 acres, the plant had more than 2 million square feet (nearly 50 acres) of floor space. E-B was then one of the largest agricultural equipment firms in the United States.

In 1937, Case purchased the floundering Rock Island Plow Company of Rock Island, Illinois, and integrated it into the Case business. Rock Island was a historic company with roots reaching way back to the early steel plows made by John Deere and others in Grand Detour, Illinois. One of the Rock Island Plow Company's preceding firms included Robert N. Tate as a partner. He had worked with John Deere in Grand Detour, Illinois, and was Deere's partner in Moline, Illinois, when Tate and Deere first began building plows there in 1848.

1944 Case LA

Four- to five-bottom plows were easily pulled by the big Model LA in 1940. The long 6-inch stroke in its 4.625-inch bore cylinders gave the LA engine extra lugging ability appreciated by belt power users. The LA 403-cubic-inch engine ran at 1,100 rpm to produce its measured 40.8 horsepower drawbar pull. Under a new rounded sheet metal grille, hood, and fenders beat the heart of the earlier Model L. A high-compression head and a new four-speed transmission updated the tractor's insides. This 1944 Model LA, owned by the Bailey Brothers of Milden, Saskatchewan, has lights and a starter. Rubber tires have been added to its cut-down wheels.

They had dissolved their partnership in 1853, and Tate began another business in 1856 with Charles Buford, building plows and other farm equipment in nearby Rock Island, Illinois. In 1882, the firm was reorganized as the Rock Island Plow Company.

By 1912, it had grown into a full-line company, offering plows, listers, planters, drills, disks, harrows, cultivators, stalk cutters, hay loaders, and even potato diggers—just about anything a farmer might need, except for tractors.

Heider Manufacturing Company at Carroll, Iowa, had an answer for Rock Island. As early as 1908, Henry J. Heider designed a light tractor using a friction drive. His Model A tractor was announced in 1911 and was soon replaced with the Model B Heider machine in 1912. By 1914, Rock Island Plow Company was selling Model Bs under the Heider name, and Heider was at work on his Model C. Still of the friction-drive design, the Model C was more conventional in other respects. It was powered by a Waukesha four-cylinder, 4.5x 6.725-inch bore-and-stroke engine, giving the tractor a four-plow rating. Ground speed was varied by moving the engine forward or back to change the effective radius of the friction plate where it contacted the engine flywheel.

1946 Case DC-4
This 1946 Case Model DC-4 has adjustable wide front axles for use with narrow row crops. Three-plow capacity kept the DC-4 ahead of growing weeds when it was introduced, while lights and a starter made it a handy tractor to drive. Case's D Series included the standard-tread Model D, the tricycle row-crop DC-3, the DH high-crop DC-4 with wide front, the orchard and vineyard DO and DV models, and the industrial DI Standard and DI Narrow, as well as a cane DCS model. Updates during the D model production from 1939 to 1955 included adding a hydraulic lift, live PTO and hydraulics, and Eagle Hitch capabilities. The FFA chapter at Escalon High School in Escalon, California, restored this tractor.

In January 1916, Heider sold its tractor business to Rock Island Plow Company, and Henry Heider worked with the firm to bring out the Heider Model D in 1917. It was a two-plow machine with a smaller Waukesha engine. Heider also developed a small cultivating tractor for Rock Island called the Model M. It came out in 1920. In 1925, the Heider 15/27 tractor was introduced.

Rock Island kept the Heider name on its tractors until 1928 when it replaced the old friction drive with a more conventional clutch and geared transmission. Then the new machines became

1940s Case VA Series brochure
This brochure introduces Case's VA Series, one- and two-plow tractors that the company said had "proved power in a smaller package."

1948 Case VAC
This 1948 Case Model VAC was the row-crop version of the VA model introduced just as World War II started. The VA models replaced the earlier V Series built from 1940 to 1942. This two-plow VAC features hydraulic implement lift and depth control, starter and lights, four-speed transmission, and 48- to 88-inch adjustable rear wheel tread. The VAH was a high-crop offering and the VAO an orchard and grove model. This VAC, owned by Glen J. Mlnarik of Howells, Nebraska, has the single front wheel option.

1954 Case VAH

Crop clearance for vegetable growers was a feature in this 1954 Model VAH high-clearance model. Adjustable front axles allowed setting the VAH up for different narrow-row and bedded vegetable crops. The VA Series of Case tractors was produced until 1955. The famous Case Eagle Hitch was available on the VA tractors in 1949. More than sixty percent of the 148,000 VA Series tractors made were VAC row-crop tractors. A. K. Gyger of Lebanon, Indiana, restored this classic tractor.

known as Rock Island tractors. Rock Island apparently discontinued tractor production about 1935. Case didn't use the Rock Island tractor design. It had its own new tractor design under development.

CLAUSEN FINALLY GETS HIS MODEL L

The introduction of the Case Model L came in 1929, after nearly five long years of development and testing. The L was had a standard-tread configuration designed with a new, inline, long-stroke valve-in-head engine. Its four-cylinder 4.625x6-inch engine powered it to 47.04 belt horsepower and 40.8 drawbar horsepower at 1,100 rpm on rubber tires in a 1928 Nebraska test. A 1929 test of the L on kerosene pegged it at 44.01 belt horsepower and 26.28 drawbar horsepower.

A pressure lubrication system as well as completely enclosed and oil-bathed drive components kept the sturdy Model L machine on the job for many years. The tractor was a 26/40, made to replace its two nearest-sized crossmotor machines. The new model came out in a new gray color, instead of the dark green of

the crossmotors. It was available on rubber tires as well as steel wheels with spades.

The Model L was almost an instant success. During its first two years of manufacture, more than 6,000 were made. In its ten-year production run, more than 34,000 came down the assembly line.

Later in 1929, the Model C debuted. Similar in most respects to the L, the Model C was smaller, rated at 17/27 horsepower and designed to pull two to three plow bottoms. Modification of the C into the Model CC (the second "C" stood for cultivator) in 1930 finally gave Case a row-crop tractor to compete with the other row-crop designs being introduced by other manufacturers. Riding the wave of the popularity of row-crop tractors in the mid-1930s, Case made nearly 49,500 Model CCs and Cs. More than 60 percent of that production were the row-crop version Model CCs.

The Model C standard-tread Case shared the same 3.875x5.5-inch four-cylinder engine with the Model CC. The two models were tested at Nebraska in 1929 and showed their official horsepower. Not surprisingly, they both made 29 belt horsepower and 17 drawbar horses when revved to 1,100 rpm on their kerosene fuel.

The CC introduced the famous "chicken perch" or fence cutter steering arm on the tricycle row-crop tractor—a controversial but familiar feature long associated with the Case row-crop tractors. The distinctive steering lever "roost," which was practical but not pretty, stayed with the line until 1955.

Longtime Case Vice President and Chief Engineer David P. Davies left his indelible imprint on the new lettered Case models

1950s Case S Series brochure

1954 Case SC-4
The Case Model SC-4 with its standard wheel tread was a two-plow tractor introduced in 1941. Case designed the S to compete with the popular Farmall H and to fill its own gap between the D and V Series tractors. The SC was basically a slightly scaled-down version of the earlier Case D. The S engine was a short-stroke 3.5x4-inch bore-and-stroke with a drawbar pull of 16.18 horsepower and belt horsepower of 21.62. The engine bore was increased to 3.375 inches late in its production, boosting its belt horsepower to 29.68. This 1954 SC-4, owned by A. K. Gyger of Lebanon, Indiana, is one of the last built. With fewer than 500 of its type made, this is a rare tractor.

1955 Case 400 Orchard
Built for ducking and running under tree branches in orchards, this 1955 Case Model 400 diesel orchard tractor looks ready to race. It's decked out in the two-tone Desert Sand and Flambeau Red colors that Case started using on its Hundred Series tractors in 1953. The Case diesel engine in the Model 400 had a displacement of 251 cubic inches with a 4x5-inch bore and stroke. At 1,500 rpm, the diesel produced 49 drawbar horsepower in Nebraska tests. The first Case diesel engine came in 1953 in the Model 500. This orchard tractor is owned by A. K Gyger of Lebanon, Indiana.

with his patented enclosed roller-chain final drive. It transmitted power from the differential to the sprockets mounted on both rear-driving axles. Case tractors used the same roller-chain drive design in models made well into the 1960s. Davies played a major role in many of the Case tractor projects, going back to the steam engine days. He was at first a draftsman at Case and helped prepare drawings for many machines built there, including the short-lived Paterson gas tractor of 1894.

NEW SHADES OF GRAY AND RED

About the time Case had its CC row-crop tractor ready to launch and its sales approach zeroed in on the row-crop competition, IHC threw a surprise pitch. IHC introduced a second-generation row-crop tractor: the simple and smaller F-12 Farmall. It had a simple design, pushing power from its small four-cylinder Waukesha engine back to its large-diameter wheels, mounted simply on single large-diameter keyed axles. Infinite row-width wheel settings were as easy as loosening a few bolts and sliding the wheels out to a measured setting. And an added bonus was that the F-12 was cheap.

Case dealers and salesmen began to complain that the only time they could sell their Case CC was when the IHC dealers were sold out of the less-expensive Farmall F-12s. Case's need for a smaller, less-expensive row-crop tractor was finally felt in Racine, and a machine was designed to compete with the little Farmall F-12.

In late 1935, the new Model RC began production as a Waukesha-powered tricycle row-crop with an over-the-top steering gear arrangement. Its rear end was the Case CC rear-drive unit. To distinguish the new machine from the more powerful (and more profitable) CC models, the new RC was painted a light gray color. Its wheels were painted red.

It was tested at Nebraska in 1936, where it pumped out 19.80 belt horsepower and 14.21 drawbar horsepower burning gasoline in its 3.25x4-inch, 991-rpm four-cylinder engine. Case management hoped the smaller RC wouldn't compete with its own larger CC tractor for sales. In fact, Case told its salespeople to back down to an RC model sale only to compete with other manufacturers' small, inexpensive row-crop tractors. They were not to sell the less-expensive RC model against their own CC.

RC production was started in Racine in late 1935 but was moved to Rock Island, Illinois, in 1937, when Case took over the Rock Island Plow Company's facilities there. Nearly 16,000 R and

1950s Case 500 Diesel advertisement
This advertisement proclaims the Case 500 diesel as the "star of all diesels," noting it was the first diesel tractor to offer power steering. It also highlights the 500's five-plow power, constant power take-off, and six-cylinder engine.

RC models were built from 1935 through the end of the R Series production in 1940. The familiar side-arm steering of the other Case row-crop tractors was added to the RC beginning in 1937. Its vertical front, over-the-top steering pedestal was then gone.

Standard four-wheel R, RI (industrial), and RO (orchard) models were added to the Case small tractor line in 1939. Only the RC was available earlier. By mid-1939, the Rs had been streamlined with a rounded cast sunburst or wheat sheaf grille, as well as a new styled hood. They also were being painted in Case's new

***Above and opposite:* 1966 Case 1200**
An all-weather cab encloses the operator's platform on this 1966 Case Model 1200 Traction King. All four wheels of the big tractor can be steered to help it into and out of tight corners. Case made the 1200 Traction King between 1964 and 1968. The Traction King led a long line of Case four-wheel-drive models. The more powerful 1470 Traction King diesel replaced the Model 1200 in 1969. The 1470 produced 132.06 drawbar horsepower in Nebraska tests. Kevin Geltmaker of Floyds Knobs, Indiana, found old combine tires to fit his Case 1200.

Flambeau Red color. Other refinements that became available on these tractors included rubber tires, lights, and a starter.

The rugged Model L standard-tread tractor had been in production since 1929 when it was finally upgraded and re-released in Flambeau Red as the Model LA in 1940. The LA featured not only a new color, but a new look with streamlined, deep, rounded rear fenders, grille, and hood.

The new five-plow tractor was hardly changed under the skin. It still boasted basically the same 4.625x6-inch bore-and-stroke engine with a slight boost in compression. Magnetos remained standard equipment on the LA long after it was offered with lights and a starter. Records show some 42,000 of the LA and its model variants were made from 1940 to 1953.

A brawny new Flambeau Red Case tractor series was launched about the same time as the lighter-duty models VC and V came aboard. In the three-plow power category long held by the Model C and CC tractors, Case launched the new streamlined D and DC models in 1939. The D and DC were basically Model CCs with a new smoothed and streamlined cast grille and metal hood.

The new grille had a rounded casting with horizontal louvers blending into a rounded sheet metal hood with a graceful rounded cutout for access to the engine. Stamped clamshell fenders added to the modern look. The gas tank also was rounded off at its rear to give the tractor operator more vision of his cultivator. A four-speed transmission gave the popular tractor working speeds from 2.5 to 10 miles per hour.

engine was a shorter-stroke, higher-rpm engine than the D's. It had a 3.5x4-inch bore and stroke that burned distillate and ran at 1,550 rpm. Tested at Nebraska in April 1941, the four-cylinder engine produced 22.20 horsepower on the belt and 19.44 on the drawbar. It was tested at a weight of 4,200 pounds.

The battery box on the SC was moved under the rear of the fuel tank, not placed on top like the D's. In 1953, an improved Model S and SC were tested at Nebraska. The 1953 version burned gasoline and had a larger 3.625x4-inch bore-and-stroke engine. Operating at a more rapid 1,600 rpm, the engine showed 31.71 belt horsepower and 27.68 drawbar horsepower. The distillate version of the new SC was also tested in 1953. It mustered a less-impressive 24.97 belt horsepower and 23.25 drawbar horses.

The S was a well-received model: 74,000 of it and its model variations were made between 1940 and 1955. Three-fourths of the Model S Series tractors made were the row-crop SC model.

SO LONG HORSES

Still doing well with its larger tractors, but experiencing some problems with sales of its smallest R Series tractor, Case reluctantly moved toward even smaller units. The demand for one- to two-plow tractors was increasing rapidly as they became the ultimate replacement for horses and mules.

Since its threshing machine and steam engine successes, Case had a long history of sales strength in the Great Plains and other extensive farming areas where small grains were grown on large operations. Just the idea of producing very small tractors went against the Case tradition, but the company knew the need for the "team-sized" tractors was great. And Case's competition was already entering that market.

Both Allis-Chalmers and IHC brought out one- to two-plow-sized machines in 1938 and 1939. A-C's Models B and C used mostly vendor components to keep down costs and were right on target for the emerging market. The Farmall A had a left-offset

Nebraska tests in 1940 showed the DC engine provided 37.28 belt horsepower and 33.06 drawbar horsepower at 1,100 rpm operating on gas. The distillate engine in the D standard-tread tractor ran faster at 1,200 rpm to produce the 35.36 horsepower measured at the belt pulley and the 30.67 on the drawbar. Both tractors weighed about 7,000 pounds when they were tested on rubber tires. Electric lights and a starter were offered as optional equipment. Both D and DC models saw some service on steel wheels when rubber shortages nearly curtailed tractor tire production during World War II.

The DC, or cultivating tractor, was confusingly referred to as the DC3 to indicate its tricycle-wheel arrangement. The other DC-3 of the era was the famous Douglas transport aircraft destined for heroic service during World War II as the military C-47 *Dakota* or *Skytrain*.

The D Series was manufactured from 1939 through 1953, with the DC machine selling more than half of the combined production for the period: just over 104,000 tractors.

Joining the DC in the row-crop lineup in 1940 was the smaller Model SC. Designed to compete with the new Farmall H, the S Series tractor was a re-engineered and scaled-down DC. Its

powertrain design for better row visibility. The Ford-Ferguson 9N, also announced in 1939, was aimed at the same market. Case jumped into the fray with its V Series models in 1940.

That year, the Case VC was introduced. It was a tricycle-type row-crop tractor powered by a Continental engine. The bore and stroke of the Continental mill were 3x4.375 inches. At 1,425 rpm on gasoline at Nebraska, the engine produced 24.48 belt horsepower and 18.35 drawbar horsepower. The engine's differential, transmission housings, and torque tube were built by Case, and the Clark Company of Michigan supplied the transmission and differential gears. Metal work was farmed out to local vendors.

The VI (industrial), V (standard), and VO (orchard) versions soon joined the VC. Priced at $625, Case's smallest tractor was finally competitive with other tractors of its size. The V Series was produced from 1940 to 1942 and turned in a production run of more than 16,000 units.

In 1942, Case replaced the V Series with the VA Series. One aim of the new model was to build more of the unit in-house and thus improve its profitability. The VA models had a Case-designed engine built by Continental, which used both Case and Continental parts. Once the Rock Island engine plant was operating in 1947, VA engines were built there. The VA's transmission and differential gearing were of Case design and manufacture.

VAC row-crop units started rolling off the line in Rock Island about a month before the bombing of Pearl Harbor, Hawaii, on December 7, 1941. By mid-July 1942, when war production closed the VA line, more than 8,500 of the little tractors had been built. During the war, only special warehouse versions of the VA came out of the plant. Once manufacturing resumed after the war's end, the Rock Island plant cranked out VAs at a rapid rate. By the end of Model VA production in 1955, more than 148,000 had been made, more than 60 percent of them being the VAC row-crop version.

1976 Case Agri-King 1570
This 1976 Case Agri-King Model 1570 *Spirit of '76* was decked out in red, white, and blue to help celebrate the nation's bicentennial in 1976. Only a few 1570s were decorated in the celebratory pattern and colors. Under the 1570's patriotic hood worked a 504-cubic-inch turbocharged diesel that was also used in the Model 2470 four-wheel-drive tractor. The two-wheel-drive 1570 produced 152.96 drawbar horsepower through its single axle via its 4.62x5-inch bore-and-stroke engine. Case specialist A. K. Gyger of Indiana practically salutes this gem.

DIESELS JOIN THE MIX

In 1953, Case introduced its first diesel tractor, the Model 500. Its six-cylinder engine put out 63.81 belt horsepower and 56.32 drawbar horsepower during Nebraska test No. 508. The Case-designed and Case-built engine used the Lanova power cell concept. Its bore and stroke was 4x5 inches with a heavily-built crank and crankcase designed to handle the stresses from the big diesel unit turning over at 1,350 rpm. Double disc brakes, electric lights, and electric starting were all standard on the Flambeau Red machine.

During its production from 1953 to 1955, a total of 5,225 of the model were built. The 500 was the last of the all–Flambeau Red tractors. New paint was on its way.

The Model 500 was only the first of several new Case model introductions in the 1950s. In early 1955, a two-tone Model 400 was introduced as the successor to the Model D. It was available in gasoline, diesel, LP, or distillate fuel versions, and in a row-crop, regular, orchard, high-clearance, or industrial configuration. Powering the new Model 400 was a four-cylinder version of the engine used in the 500. It turned over at 1,450 rpm, or another 100 rpm as compared with the 500 engine. Engine and castings on the new model series were painted in Flambeau Red with its sheet metal a contrasting Desert Sunset color.

Soon joining the two-tone fleet were the Model 300, which replaced the S Series in 1956, and the Rock Island–plant made Model 200. The Model 200 replaced the VA models. In 1956, the 500 was upgraded to the Model 600 with the addition of new styled grilles and the two-tone paint previously used on the less-powerful models.

THE CASE FOR CRAWLERS

Case got into the track-laying tractor market in 1957 when it bought the American Tractor Corporation of Churubusco, Indiana. The firm's Terratrac models included GT-25 and GT-30 gasoline models and a DT-34 diesel. Hydraulic lifts, power take-offs, and six-track gauges from 36 to 72 inches for different row spacings were available on the Terratrac.

Case continued production of the track-type machines at Churubusco through the 1950s. In 1961, their manufacture was shifted to the Case facilities in Burlington, Iowa, and the crawlers became the foundation of the Case industrial equipment line.

Left and opposite: **1976 Case Agri-King 1570**
This *Spirit of '76* Case 1570 regularly is a part of Fourth of July celebrations. Its heavy-duty three-point hitch is capable of mounting big implements.

AN UNLIKELY PAIRING

In the 1960s, The J. I. Case Company was about to undergo many more changes, including the eventual mind-blowing reality that Case would survive over industry-leader International Harvester Company. Sizewise, they weren't in the same league, but that's what happened.

Case's ownership changed over the years, and in 1967, Kern County Land Company, a majority Case stockholder, was bought by Tenneco of Houston, Texas. Tenneco subsequently bought International Harvester Company in 1984 after the farm equipment industry giant suffered massive losses due to the farm crisis of the 1980s. Tenneco restructured in 1988, and the joined Case-IH operation became its largest division. In 1994, Tenneco spun off the Case Corporation as a new company. Its products are now identified as Case-IH machines.

In 1999, Case Corporation merged with New Holland N.V. The new group is known as CNH Global N.V. of Amsterdam. The Case New Holland names inspired the CNH initials. CNH ranks as the second-largest agricultural equipment maker behind Deere & Company with 2005 worldwide sales of $12.575 billion.

CNH is regionally organized with separate brand-driven commercial organizations and distribution networks. It currently operates in North America, Europe, and Brazil. Many of its brands are familiar North American historic brand names, including Case, Ford, International Harvester, New Holland, and others from around the world.

Case-IH still designs, manufactures, and sells its line of red tractors. Its largest and most impressive current tractor is the huge Case-IH Steiger Quadtrac machine. The largest of the Quadtrac models, the STX530, harnesses 530 horsepower from its 15-liter diesel engine. It has four rubber track drives that tiptoe across farm fields, applying only 9.73 pounds per square inch of pressure to the soil.

Case-IH offers a full range of wheel tractors—everything from compact models ranging from 19 to 37 horsepower and utility models from 46 to 80 horsepower to higher horsepower machines with 85 horsepower on up through the Quadtrac's 530 horsepower capability.

Chapter Four

Caterpillar

1926 Caterpillar 2-Ton

Tackling farm tillage and other heavy pulling chores was the forte of this small 1926 Caterpillar 2-Ton tractor. Originally introduced in 1923 as a design of Holt Manufacturing Company of Stockton, California, the 2-Ton model survived the 1925 merger of Holt and the C. L. Best Company. The 2-Ton engine was an advanced four-cylinder gas-burning overhead-cam valve-in-head engine. Its 4x5.5-inch bore and stroke produced 15 drawbar horsepower and 25 belt horsepower. Light gray paint with red lettering marks this machine owned by Bill Walter of Missouri.

Tractors that "crawled" over soft farm fields grew up in the American West, where soil conditions often called for more flotation than wheeled tractors could muster. The Holt Manufacturing Company of Stockton, California, tested its first track-type machine in 1904, when long tracks with two-foot-wide wooden treads were attached to replace the rear driving wheels on a Holt steam engine.

Earlier, Holt and others tried to "float" their heavy machines on the soft delta soils of the San Joaquin River delta near Stockton. They attached huge wooden rim extensions, sometimes as wide as 18 feet, on steam engine drive wheels in an attempt to keep the monster machines from sinking into the peat soils. Sometimes that worked and sometimes the steam-powered wheel tractors mired in the muck. Farmers could spend days extricating them.

The endless track with its wide tread worked, and the continuous track idea was adopted. While watching the moving tracks on one of the early machines, the track contacting the ground appeared stationary and the advancing track galloping forward on top was in motion. One observer, photographer Charles Clements, noted, "It crawls, just like a caterpillar." Holt President Benjamin Holt agreed, and Caterpillar was registered as a trademark for the track-laying machines in 1910 by the Holt Manufacturing Company.

Holt sold its first steam crawler tractor in 1906 for $5,500 to a Lockport, Louisiana, land company. The Louisiana customer had earlier buried a Holt steam wheel tractor as it attempted to plow Mississippi Delta land. The steam crawler tractor that replaced the earlier wheel-type engine was used successfully on the marshy Louisiana ground for many years.

The first Holt gasoline-powered crawler tractor was built in 1906. In a trial by fire, Holt sold twenty-eight of its new gasoline crawler tractors to the City of Los Angeles, California, to help transport construction materials on the city's massive 230-mile-long aqueduct project during 1908 and 1909. The $23-million project was built to bring water across the Mojave Desert to the city from the Sierra Nevada Mountains.

Working in tough unpaved conditions—in desert sand, rocks, and even snow—the tractor's weaknesses showed up fast. Holt had to scurry to keep them running. The firm's proving ground experience showed it how to build better tractors—those with all-steel construction, better spring suspensions, improved clutches, three-speed transmissions, and enhanced overall durability.

In an expansion move in 1908, Holt bought out Daniel Best Agricultural Works of San Leandro, California, its largest competitor. Daniel Best's company was founded in 1885 and had competed with Holt for many years in building steam traction engines and combines. Best was already in his seventies and wanted to retire. For a reported sum of $325,000, he turned his business over to Holt.

Above and opposite: 1913 Caterpillar 60
Crawling their way over mud and soft earth, early tracklayer tractors were soon called Caterpillars. This 1913 Model 60 Caterpillar was made by the Holt Manufacturing Company of Stockton, California, and was carried by a single heavy-duty front wheel. Its 60-horsepower engine, riding well forward on the frame, was soon moved further back over its tracks, after Holt engineers figured out that the front-mounted tiller steering arrangement was not needed in 1921. Holt registered the Caterpillar trademark in 1910. The Holt 60 has a 1,230-cubic-inch valve-in-head engine. Larry Maasdam of Clarion, Iowa, and Ron Miller of Michigan share ownership of this rare California machine.

Under the terms of the Holt-Best agreement, Daniel Best's son, C. L. "Leo" Best, would buy a third of the pre-sale Holt stock. He would then be a substantial Holt stockholder. For whatever reasons, he didn't exercise that option.

Leo did stay on at the company as an employee, working with Holt in Stockton for only two years after the 1908 Holt-Best merger (a sale Leo had originally opposed). In 1910, Leo Best, with financial help from his father, left Holt and struck out on his own as the C. L. Best Gas Traction Company of Elmhurst, California. The long-term Holt-Best competition was far from over.

The new Best company first offered gasoline wheel tractors in 60- and 80-horsepower sizes before introducing its first crawler tractor, the 17-ton Tracklayer Model 75, in 1912. The Tracklayer name became the trademark of Best's crawler machines after Holt protested when Best first tried to use the Caterpillar designation.

Although the Best Model 75 looked just like the Holt Caterpillar Model 75, Best had improved the tractor significantly, especially in the use of improved metals that added useful life to many wear-prone parts.

On December 31, 1912, in an effort to put its own house in order and build a firm footing for future operations and expansion, the Holt interests consolidated its eight different Holt companies into one overall outfit, named Holt Manufacturing Company.

Meanwhile, C. L. Best continued to develop and market crawler tractor models. Soon both Best and Holt realized that the front tiller wheel was not needed on the crawler tractors. In 1915, Best made his new 30-horsepower Muley and 16-horsepower Pony Tracklayer tractors without the big front tiller wheel. At about the same time, Holt dropped that feature from its line of Caterpillar tractors. The redesign put all of the crawler's weight on the tracks for improved traction.

In 1916, when he needed more manufacturing space, C. L. Best bought back his father's old tractor factory facilities in San Leandro, California, and transferred manufacturing there.

WHEAT BACK EAST

Vast new areas of wheat farms were being developed in the Canadian prairies and the U.S. Great Plains as the twentieth century got underway—waving fields of grain that looked like a huge and expanding market to Holt for his company's gas crawler tractors and combines. In 1909, a new Holt outfit, the Northern Holt Company, was organized and headquartered in Minneapolis, Minnesota, to take advantage of the new "eastern" markets. The new company assembled two tractors in Minnesota before it

1920s C. L. Best 60 Tracklayer
Daniel Best's son, C. L. Best, worked with Holt for two years after 1908 merger. But in 1910, Best formed his own company, building gas wheel and track tractors of 60 horsepower and 80 horsepower.

ran short of operating funds. It was at this juncture that Holt learned of a large factory in East Peoria, Illinois, that was vacant and available.

Holt, with help from a friendly local implement dealer, Murray M. Baker, negotiated the purchase of the Illinois plant for $75,000, with just $11,500 down. The factory was worth some $250,000. The Colean Manufacturing plant became the Holt Caterpillar Company facilities in January 1910, with Baker serving as general manager. Colean had manufactured steam engines and threshing machines in its East Peoria plant, but it failed as a rapid transition from steam to gas tractors and combines left it behind.

WARTIME CHANGES

The Caterpillar tractors Holt sold to the U.S. Army for service in World War I won the company many accolades. The tough machines plowed their way to the war front in Europe, overcoming fields of mud and pulling wagon trains loaded with critical supplies. Other Army Cat tractors had armor plating installed and were used to tow heavy field artillery into battlefront firing positions.

The Caterpillar crawler undercarriage inspired the British-built tanks that first rolled over German trenches in the Battle of the Somme in France in 1916. That successful use of a track-type undercarriage on armored vehicles changed warfare forever. Britain's acclaimed champion of the tank concept, British Army

1928 Caterpillar 2-Ton

Louvered metal side curtains still shield the engine compartment on this 1928 Caterpillar 2-Ton. This model's lever-operated steering clutches, with help from brakes on each track, stop one track while the other track turns the crawler its own length. Highway Yellow paint replaced the gray paint of former Caterpillar models in 1931. This tractor is still on the same Illinois farm where it started working as a new tractor nearly eighty years ago. It is owned by Robert Garwood of Stonington, Illinois.

An Early C. L. Best Tractor Company advertisement

1929 Caterpillar Fifteen
The Caterpillar Model Fifteen was a step up in power from the smaller Model Ten Caterpillar. The Fifteen turned out 25 belt horsepower from its four-cylinder 200-cubic-inch engine. This 1929 model is considered the "big" Fifteen and is owned by Richard Opdahl of Alden, Minnesota. A 1932 "small" Fifteen, with a less powerful engine, later replaced the Model Ten. The optional canopy with canvas side curtains on this Caterpillar gave the operator some comfort during early spring or late fall tillage operations.

Caterpillar Magazine cover
Issue number thirty-one of Caterpillar's company magazine featured a Model Twenty, which was introduced in 1928 with a 20/35 horsepower rating. The Twenty had a four-cylinder 4x5.5-inch bore-and-stroke engine of 277-cubic-inch displacement that produced 29 belt horsepower.

Colonel Earnest D. Swinton, dubbed Holt's Stockton, California, works as "the cradle of the tank."

With rapid growth in both numbers of models and sales volume, C. L. Best's growing competition became a concern for Holt after World War I. During the war, Holt had concentrated on making military tractors for the government, and at the armistice in 1918, the company suddenly found itself with no military contracts and with too few farm-sized tractors ready for developing gas farm tractor markets.

Fortunately for both Holt and Best, C. L. Best had concentrated on designing, building, and selling mostly farm-sized tractor models during the war years. Holt now needed these machines in its line. Best had not fared well during the war in selling equipment to the military and needed more markets and manufacturing facilities. So a "consolidation" of the two firms made sense to both parties.

On March 2, 1925, the C. L. Best Gas Tractor Company became a part of the newly formed Caterpillar Tractor Company. The new company combined the assets, patents, and ideas of the Holt and Best crawler lines. C. L. Best became Caterpillar's first chairman, and the new company's headquarters was located at the

1929 Caterpillar Twenty
The Caterpillar Model Twenty "step up" starter is this hand crank. By just stepping up to it, an operator can start the Cat.

former Best tractor plant in San Leandro. Both firms had survived tight financial times brought on by a recession following World War I. Many other firms in the farm tractor business didn't fare as well and were forced out of business.

PEORIA TRACTORS

Following the organization of the Caterpillar Tractor Company in 1925, manufacture of tractor models from both the Best and Holt lines was moved to Peoria, and a Caterpillar subsidiary, the Western Harvester Company, was set up at Stockton to continue manufacturing Holt combines. The combined harvester had long been a mainstay of the Holt product line. Holt began making its first horse-drawn combines in 1886, just a year after Best made its first combine in 1885. In the forty-three years between 1886 and 1929, Holt made and sold more than 14,000 of its combined harvester. Combine production was later moved from Stockton to Peoria, where Caterpillar combines were made during the Depression in the 1930s.

In 1925, the Peoria plant started turning out Caterpillar versions of the Best 30- and 60-horsepower models. Prior to the merger, the Holt models included the 2-Ton, the 5-Ton, and the 10-Ton. The 2-Ton was introduced in 1923 and was a small state-of-the-art tractor with a Holt-made four-cylinder overhead-valve engine. The little 2-Ton was a 25-belt horsepower, two-three plow machine. After the merger, the 5-Ton and 10-Ton Holts were dropped from the line, but the 2-Ton was continued. Caterpillar pursued the agricultural tractor market in the 1920s and 1930s with models ranging in power from two- to six-plow ratings.

In 1928, the new Caterpillar Model Twenty, with a 20/35-horsepower rating, was tested at Nebraska. In 1929, the Caterpillar Model Fifteen was announced. With those two new models, Caterpillar had five models: the Fifteen, Twenty, Thirty, and Sixty. The model designations roughly correlated to their drawbar horsepower ratings, not their weight as previous crawler designations indicated. A high-clearance version of a Model Ten was unveiled in 1931.

DECADE OF DIESELS

Caterpillar's big news in 1931 was the introduction of its Diesel Sixty-Five, the first agricultural diesel tractor offered for sale in the United States. Before Caterpillar's pioneering work on its new

1932 Caterpillar Ten

Smallest of the Caterpillar tractors was the Model Ten. This little 1932 Model Ten is equipped with optional rubber track pads for working on pavement or turf. It has a belt pulley and electric lights, but no starter. Its four-cylinder side-valve engine has four 3.375x4-inch bore-and-stroke cylinders displacing 143 cubic inches. The Model Ten was a 10/15 tractor built at the Peoria factory from 1928 to 1932. When collector-restorer Dwight Pletcher of Goshen, Indiana, found this tractor, it was equipped with a front blade for snow removal.

1937 Caterpillar Twenty-Two Orchard

The mission of this 1937 stealth Caterpillar Model Twenty-Two Orchard was to sneak through orchard and grove trees without damaging them. The Model Twenty-Two was the most successful of the smaller Caterpillar models with more than 15,000 manufactured from 1934 to 1939. The Model Twenty-Two had a new 4x5-inch bore-and-stroke overhead-valve engine with 251 cubic inches of displacement when it was first built in 1934. It was a 25-drawbar horsepower tractor. Larry Maasdam of Clarion, Iowa, owns this Cat tractor.

diesel engine, diesels had been used mostly for huge marine or stationary engine installations.

The Diesel Sixty-Five's engine was a four-cylinder four-stroke mill turning at 650 rpm. Starting the diesel was achieved by first starting an interconnected two-cylinder gasoline engine. The small pony engine was then clutched into the diesel engine to turn it over and get it running on its own power. German engineer Rudolph Diesel first worked on the diesel engine design in 1892 while still a student. Within a relatively few years, thousands of stationary diesel engines were hammering away in installations that required efficient long-running power at constant loads and speeds.

The diesel engine is typically a four-cycle engine that substitutes the heat produced in the compression stroke for a spark plug or other ignition device. Fuel is injected under pressure into the combustion chamber at the top of the compression stroke. Upon compression, the air in the cylinder is hot enough to ignite the fuel, ignition occurs, and the power stroke results. Compression ratios in diesel engines are extra high to create the needed heat for combustion. With the high oxygen density created by the extra high compression in the combustion chamber, the diesel engine is able to extract maximum energy from higher density fuels. Typically a diesel engine saves about 50 percent on the fuel burned compared with gasoline. The diesel tradeoff is that diesel engines have to be heavier and stronger than gasoline engines of the same power size, just to survive the extra forces at work in the diesel combustion chambers and crankshaft. The injection pumps that squirt the fuel into the compressed hot air in the combustion chamber have to be extremely precise as to timing and the amount of fuel metered,

1937 Caterpillar Twenty-Two Orchard
Above: The operator's seat and controls on the orchard Twenty-Two were lowered and moved to the rear to get them out of tree limb range. *Right:* The crawler's hand clutch, steering clutches, brakes, gear shift, and throttle were moved back with operating rods to be accessible from the seat. Metal shields kept tree branches from getting caught in the Twenty-Two's tracks.

and they have to be strong enough to overcome the high compression inside the cylinder. Add a starting device to get the engine rolling so it can "diesel" on its own and there's another complication. Those extras all add to the cost of the machine.

Caterpillar put years of research into designing its diesel engines, efforts that were well rewarded. By November 1935, Caterpillar had made 10,000 diesel engines. The company was able to sell its tractors going into the 1930s because of their diesel engines and their proven fuel efficiency, though times were tough economically.

In 1932, Caterpillar brought out its Model Twenty-Five. In 1935, three new diesel-powered crawlers and another gasoline tractor were announced. The new diesel models were the RD8 with 95 drawbar horsepower, the 70-horsepower RD7, and the 45-horsepower RD6. The gasoline-powered tractor was the 30-horsepower R4. A diesel version of the R4, designated the RD4, came along in 1936. It weighed 10,100 pounds, and its four-cylinder diesel turned at 1,400 rpm.

Cat's smallest diesel, the D2, arrived in 1938. It was designed for farm work and could pull three to four plows or a 10-foot disk harrow. Nebraska tests showed drawbar and belt horsepower on the D2 at 19.4 and 27.9, respectively. The D2 weighed 7,420 pounds. Its gasoline and distillate burning counterpart was the R2, which was a bit lighter at 6,835 pounds, but tested at similar horsepower output. The D8's horsepower was increased in 1940, as was the D7's. A six-cylinder D6 replaced a smaller D6 in 1941.

The big Cat crawlers used for farming were primarily utilized on large acreages where gang hitches of tillage or planting equipment that covered wide swaths made them very efficient. Western wheat growers welcomed the crawler tractors' extra size, weight, and power during tillage and planting. Their size and power was needed again at harvest to negotiate steep slopes with their big pull-type side hill combines. Midwestern farmers gravitated toward the smaller Caterpillars and used them for primary tilling, preparing seed beds, and pulling heavy combines and corn pickers on sticky black "gumbo" soils.

Diesel-powered farm tractors were developed by other tractor makers with increasing frequency after World War II. Caterpillar led the way in the 1930s, but in 1941 the company went off to war again and, in a sense, didn't make it back to the farm after that. Caterpillar bulldozers, scrapers, and other machines played a huge role in the war effort at home and in both European and Pacific theatres. The publicity Caterpillar received for the accomplishments of its machines proved invaluable after the war, as a stepping stone to an increasing use of Cats in construction applications. Post-war there was a lot of work to be done in the United States, and the Cats were soon there to do it.

From 1930 forward, row-crop tractors became the dominant farm tractor type. Equipped with rubber tires and more powerful, high-compression gasoline engines in the mid-1930s, the larger row-crop machines gave farmers

added utility. They were useful for plowing and disking seed beds, as well as cultivating crops.

By the mid-1950s, diesel engines were showing up in row-crop tractors from many farm tractor makers. Next came four-wheel-drive tractors to do more and more of the heavy pulling on U.S. farms.

CATERPILLAR CHALLENGER LINE

Caterpillar seemed reluctant to rejoin the competition in the agricultural tractor market well after World War II. It wasn't until 1986 that Caterpillar announced a new rubber-belted farm

1930s Caterpillar advertisement
In this Caterpillar advertisement, the company isn't just touting its Caterpillar Thirty but the whole line of tractors, which promise "the day's work done in time for an evening of fun." It does list the Thirty's price as $3,000, though.

tractor series aimed to compete with four-wheel-drive rubber-tired tractors.

The Challenger Series Model 65 ran on 24.5-inch-wide steel cable-reinforced rubber belts instead of steel tracks. Caterpillar called its new rubber-track undercarriage the Mobil-trac system. It was built to handle higher field speeds and over-the-road transport at more than 18 miles per hour. Powered with a 638-cubic-inch six-cylinder turbocharged and intercooled diesel, the Model 65 had a powershift transmission with ten forward speeds. A modern cab shielded the operator from the weather and provided a comfortable workstation.

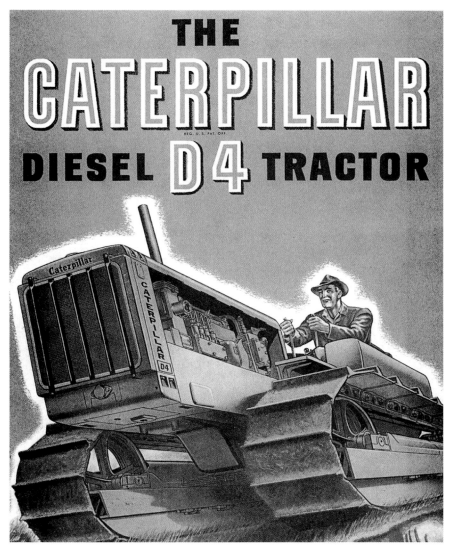

Cat Diesel D4 advertisement
In 1931, Caterpillar was the first tractor manufacturer to offer a diesel tractor in the United States. Over the years, the company continued to invest money in diesel tractor development, which paid off as several of its diesel models were extremely popular. It introduced the D4 diesel tractor in 1938, which put out 50 horsepower on the drawbar and 59 horsepower on the belt.

In 1991, a 15-horsepower boost in the Challenger engine made it into the 250-PTO-horsepower Model 65C. Cat next added a Challenger Model 75 with wider tracks and 280 PTO horsepower in 1991. In 1994, the Challenger line was expanded to four models: the 65C, 70C, 75C, and 85C. The three largest models ran on 30-inch-wide tracks and had up to 305 PTO horsepower.

In 1995, Caterpillar introduced another new concept for crawler-type tractors: the Challenger 35 and Challenger 45 rubber-belted track models, specifically designed for row-crop use. The tracks of the two new crawler row-crop models were adjustable from 60 to 88 inches in width to match crop-row widths. The 35 and 45 were rated at 175 and 200 horsepower, respectively, and had Caterpillar 403-cubic-inch engines. They had 16 forward and 9 reverse speeds from their powershift transmissions.

Caterpillar added a big new combine to its agricultural product line in 1997. A joint venture between Caterpillar Inc. and Claas of Germany introduced the Lexion combine to North American farmers by way of Caterpillar dealers. The Lexion, powered with Caterpillar diesel engines, was available with or without Caterpillar's Mobil-trac System, a rubber-belted track undercarriage. A new factory was built in Omaha, Nebraska, to manufacture the combines. By September 2001, the plant was busy making the Lexion combines.

Cat was now back in farm fields in an impressive way with new modern, cutting-edge products.

CATERPILLAR TODAY

Today, Caterpillar produces a long line of industrial products, including the company's well-known crawler tractors and bulldozers, as well as huge off-road trucks, backhoes, track loaders, wheel loaders, excavators, lift trucks, motor graders, tree skidders, highway pavers, profilers, and more. Caterpillar is the world's leader in diesel engine production. In 1996, it purchased diesel maker MaK Motoren of Germany, and in 1997 it added Perkins Engines of the United Kingdom to its diesel-engine stable.

In 2002, Claas bought back Caterpillar's 50 percent share in the joint combine venture. At the same time, Caterpillar dissolved the other joint venture with Claas to

Caterpillar Thirty-Five Diesel
The Cat Thirty-Five diesel was one of the company's earliest diesel crawlers, introduced in 1933. Its 5.25x8-inch bore-and-stroke engine had a 520-cubic-inch displacement and revved at 850 rpm. The Thirty-Five was rated at 39 drawbar horsepower and 45 belt horsepower.

build Challenger rubber-belted tractors for sale by Claas in European farm markets. Caterpillar dissolved the marketing pact with Claas in light of an agreement with AGCO of Georgia to sell AGCO the design, assembly, and marketing rights to the Challenger line of rubber-belted crawlers.

AGCO, whose roots trace back to the Allis-Chalmers Company of West Allis, Wisconsin, is currently selling the Challenger line through participating Caterpillar dealerships. AGCO is headquartered in Duluth, Georgia, and plans to introduce

a 500-horsepower, rubber-tired, four-wheel-drive Challenger Cat tractor in 2007.

Meanwhile, Caterpillar, Inc., based in Peoria, Illinois, continues to grow. In 2005, Cat was the largest maker of construction and mining equipment, diesel and natural gas engines, and industrial gas turbines in the world. In fiscal 2005, Cat reported sales and revenues of $36.34 billion from its worldwide operations. Slightly more than half of that total (53 percent) comes from the company's North American operations.

1995 Caterpillar Challenger

Considered the ultimate farm tractor when introduced in 1995, the Caterpillar Challenger 55 row-crop came in three different power sizes. This is the 1996 Challenger 55 with a 200-horsepower rating. Challenger rubber tracks are sized and axles adjustable so they can work in a variety of row-crop spacings. In 2002, Caterpillar sold manufacturing and marketing rights of the Challenger line to AGCO of Duluth, Georgia. Daniel Machens of Portage des Sioux, Missouri, owns this powerhouse Cat.

Chapter Five

John Deere

1936 Deere BO
Fairings on the Model BO Orchard hood protect the fuel filler caps and the air inlet. The BO's muffler exhausts are down from the engine, as on the BR and BI models. The rear fenders have extra sheet metal in front to keep interfering branches at bay. The driver's seat and controls are also lowered to help keep the operator out of the limbs. Bruce Wilhelm of Avondale, Pennsylvania, restored and owns this 1936 Model BO orchard, which he dolled up with shiny brass lug nuts.

J ohn Deere didn't give the world the steel self-scouring plow, but he did make it widely available. Deere, a thirty two-year-old black-smith from Rutland, Vermont, made his way west to the pioneer village of Grand Detour, Illinois, in late 1836, drawn there by the promise of new opportunities, which he soon found.

John Deere came to Illinois with very little more than his black-smithing tools, but was soon working his trade—mending farm equipment, wagons, and stage coaches; making hand tools; and doing whatever else he could do in the small frontier town. His wife and children were still back East, waiting to see if John's venture would be successful.

Grand Detour, the village where Deere settled, was named for its location on a picturesque loop in the Rock River. It was surrounded by vast prairies of mixed tall native grasses. New settlers were just discovering that under that heavy mat of grass and prairie plants were deep fertile soils. One exaggerated tale of the time about the soil's fertility was that if you stuck a crowbar into the black soil at night it would sprout ten penny nails by morning. Previously, some had mistakenly believed that if the prairie couldn't grow trees, it couldn't grow crops.

The prairie soil was indeed fertile, but not easy. Breaking through the sod was the first challenge for settlers. Keeping the heavy deep soils cultivated was an unending task. That's where Deere and his products came in. His new hometown was in the midst of a sea of prairie . . . and opportunity.

STICKY GUMBO

Much of the sod breaking was hired out as custom work to operators using huge (by the standards of the day) cart-mounted, 16- to 24-inch, single-bottom sod-breaking plows, which were pulled by four to six plodding oxen. Some farmers did their own sod breaking with smaller plows and smaller hitches. Breaking, or turning the heavy sod, allowed the vegetation to deteriorate so cropping could start.

It was in the years following the sod breaking that the new prairie farmers ran into a sticky problem. The heavy black gumbo soil stuck to their cast-iron plowshares and moldboards. Gummed up with sticky soil, the moldboards wouldn't turn the soil, but merely pushed it aside, with the plow coming out of the ground in the process. No matter how often the farmers scraped and scoured the bottoms with paddles or blades, the soil would cling again just as soon as the plow went back into the ground. Plowing in those conditions was painfully slow, if not impossible.

Deere recognized that there was a need for plowshare and moldboard that would polish to a smooth surface and "scour" itself, so the heavy soils would slide by as the plow turned the furrow. In about 1837, Deere took part of a broken steel sawmill blade, cut off the teeth with a chisel, heated it, and hammered it into a plow, then offered it for use. The steel, probably from a vertical, straight, jigsaw-type saw of the era, was already polished from its up and down strokes cutting Ogle County oak.

The plow worked. It scoured on its own and turned over the black soil in long greasy ribbons. The good news spread to neighboring farms and Deere was soon busy hammering and polishing plows, first as a solitary venture, then with a series of partners. John Deere later remembered making two plows in 1838, ten more in 1839, forty in 1840, seventy-five in 1841, and one hundred in 1842.

PLOW PARTNERS

Deere and others must have heard opportunity knocking. Leonard Andrus, owner of the local sawmill where Deere probably found his broken sawmill blade, partnered with Deere in early 1843. That partnership operated as L. Andrus & Company. They built a two-story Plough Manufactory on the Rock River near Deere's blacksmith shop and got serious about plow making. In

1844, another partner joined them, a local merchant, Horace Paine. By 1846, Paine was gone and Andrus and Deere took in Oramil C. Lathrop as a partner. They operated as Andrus, Deere & Lathrop until June 22, 1847, when Lathrop apparently dropped out and the partnership became Andrus & Deere. By then, the firm was making 1,000 plows per year. The next year brought even more changes.

In May 1848, the Andrus & Deere partnership dissolved and Deere moved downriver near the junction of the Rock River and the Mississippi River. Deere formed a new partnership in Moline, Illinois, with Robert N. Tate, an Englishman who had installed steam power in the Grand Detour "manufactory" and had worked

***Above and opposite:* 1916 Waterloo Boy R**
This 1916 Waterloo Boy Model R was built before John Deere bought the Waterloo Gasoline Engine Company in 1918. Chain-wrap steering and a one-speed transmission characterized the Model R, which was first introduced in 1914. The Waterloo Boy's two-cylinder, side-by-side horizontal engine set the direction of later John Deere engine designs. It had a bore and stroke of 6.5x7 inches. The Waterloo Boy R operator had this spring-mounted steel pan seat (above) for "comfort." With steel lugged wheels, the generous platform provided a place to stand for relief from the jolts. This classic tractor is owned by Kent Kaster of Shelbyville, Indiana.

there with Deere and Andrus. Moline, then a mere village on the Mississippi, held more promise as a manufacturing site than Grand Detour did. More waterpower and better river transportation were powerful incentives for the move.

The Deere & Tate partnership was already at work making plows in Moline by September 1848. John Gould, a businessman friend of Deere from Grand Detour, was signed as another partner that fall, and the new Moline firm became Deere, Tate, & Gould. By 1852, the group was making 4,000 plows a year and had added a

grain drill to the line. But soon the partnership members went their separate ways. Tate became a competitor in 1856 when he and Charles Buford, along with Buford's son Bassett, formed a full-line plow company called Buford and Tate. That business was the basis of what later became the Rock Island Plow Company.

Back in Grand Detour, former Deere partner Leonard Andrus was still making plows with his brother-in-law, Colonel Amos Bosworth II. The Andrus firm continued with other partners and eventually became the Grand Detour Plow Company of

1924 Waterloo Boy N
By 1924, when John Deere built this late-model Waterloo Boy N, automotive-type steering and a two-speed transmission had been added. Deere engineers had also replaced bolts with rivets to stiffen the tractor's frame. This was the next to last Waterloo Boy tractor made. The Model N was the first tractor tested at the new Nebraska Test Station in 1920. It produced 12.1 drawbar horsepower and 25.51 belt horsepower there. Jim Russell of Oblong, Illinois, owns this vintage tractor.

Deere Plow Works
Deere & Company was at first resistant to try its hand at making tractors, instead focusing its efforts on producing the plows that made the company famous. Here are a few of the company's employees at the John Deere Plow Works in 1900, more than twenty years before Deere developed a tractor of its own.

nearby Dixon, Illinois. The J. I. Case Threshing Machine Company of Racine, Wisconsin, purchased the Grand Detour Plow Company in 1919, thus giving the Case company some claim to the early Illinois steel plow history.

JOINED BY FAMILY

Operating simply as "John Deere" between 1853 and 1857, the firm's plow production more than tripled to 13,400 units. By 1854, John Deere's son, Charles, was working with the company. Following reorganizations in 1858, brought on by cash-flow problems, Charles became responsible for management. John Deere, fifty-eight at the time, remained personally involved with the firm, but apparently played a reduced role in management. Still growing in the 1860s, the company introduced the Hawkeye sulky corn cultivator, the first Deere machine with a seat, in 1863. The famous Deere-made Gilpin sulky plow of the 1870s was based on the earlier sulky cultivator.

Deere & Company incorporated in 1868, ending the long parade of John Deere partnerships tracing back to the early days at Grand Detour. By that time, the company was selling more than 41,000 plows, harrows, and cultivators annually. The Gilpin sulky plow was a welcome addition to the line, and by 1883, its sales numbers combined with the sale of more spring cultivators, harrows, and shovel plows put company sales at 100,000 units per year.

John Deere died in 1886 at the age of eighty-two, almost fifty years after he arrived in Illinois. Historians credit him with seeing value in new ideas and adapting them to his products to better serve his market and customers. Rather than being a major inventor, he is considered an adapter and an imaginative marketer.

EXPANDING THE PRODUCT LINE

Deere & Company and its branch houses spent the next twenty-five years producing more than just tillage tools—its prior niche. As the twentieth century dawned, Deere & Company made a careful investigation of farm equipment and the manufacturers that could complement its strength in tillage tools. Then it moved on its findings.

1925 Deere "Spoker" D
The first of the new John Deere–designed tractors that eventually replaced the Waterloo Boys was the Model D of 1923. This "Spoker" Model D was made in 1925 and has a 24-inch open-spoke flywheel that replaced the original 26-inch flywheel. Spoked flywheels were used on the D until 1926 when it was changed to a solid flywheel. Deere advertised the D as a 15/27 machine that was "dependable, economical, and durable." A 1924 Nebraska test pegged the D horsepower at 22/30. This Model D "Spoker" was restored by Ken and Bob Burden of New London, Iowa.

1953 Deere D Rice Special

The Model D was finally styled in 1939. This 1953 Model D Rice Special is representative of the last of the D model made. Called "streeters" by today's collectors, these tractors were assembled outside the Waterloo, Iowa, plant on a company street in the summer of 1953. Tested on rubber tires in 1940, the Model D, with its 6.75x7-inch bore-and-stroke engine, provided 38.02 drawbar horsepower and 42.05 belt horsepower. Mike Williams of Clinton, Iowa, owns this Model D.

1937 Deere D

This 1937 Model D tractor was painted gold by the John Deere Kansas City Branch to mark the one hundreth anniversary of Deere & Company's 1837 founding. Rare tractor collector, restorer, and detective Charles Q. English Sr. of Evansville, Indiana, learned of this gold tractor's existence, traced it down, bought it, and restored it back to its splendor. By 1937, the Model D was equipped with a three-speed transmission and had been tested at 30.74 drawbar horsepower and 41.59 belt horsepower. The original "Johnny Popper" was growing up.

Previous spread: 1928 Deere GP
The first John Deere row-crop tractor was the General Purpose or Model GP tractor. It was built to straddle one row in a three-row configuration. It offered power farming from four sources: the drawbar, belt pulley, power take-off, and a power lift for raising mounted implements. It was initially called the Model C. In the GP's 1928 Nebraska tests, the 5.75x6-inch bore-and-stroke kerosene engine put out 17.24 drawbar horsepower and 24.97 belt horsepower. This 1928 GP is in the collection of retired Deere ad man Don Huber of Moline, Illinois.

1931 Deere GP

A mounted three-row planter on this 1931 Model GP positioned the planter units behind the tractor with one row centered between the wheel tracks and the other two rows outside the tracks. Planter wire for check-row planting of corn was carried on the reel at left front. In 1931, the GP engine had its stroke lengthened to 6 inches and its speed boosted to 950 rpm to deliver 18.86 drawbar horsepower and 25.36 belt horsepower when tested on distillate fuel at Nebraska. Wayne Bourgeois of Kahoka, Missouri, collected this GP and planter.

1930 Deere GPP (Potato Special)
Potato growers got their own specialized tricycle-type row-crop tractor from John Deere in 1930. The Model GPP had a rear tread width of 68 inches to cultivate two 34-inch potato rows. A front frame mount for the cultivator positioned it where the operator could accurately steer it down the rows. The operator sat on an adjustable seat behind the rear axle. It could be positioned left or right for the best view of the cultivator. Wayne Bourgeois of Kahoka, Missouri, collected and restored this rare GPP.

Between 1907 and 1912, Deere added the Fort Smith Wagon Company under its wing; formed John Deere Plow Company Ltd. in Canada; organized the John Deere Export Company in New York City; bought the Marseille Manufacturing Company of Marseille, Illinois, for its corn shellers and elevators; added Kemp & Burpee of Syracuse, New York, for its line of manure spreaders; took over the Dain Manufacturing Company of Ottumwa, Iowa, and Welland, Ontario, for its haymaking tools; picked up the Deere & Mansur corn planters, the Syracuse Chilled Plow Company, Union Malleable Iron,

and Reliance Buggy Company; bought Van Brunt of Horicon, Wisconsin; put the John Deere Wagon Works under Deere & Company; designed and built Deere's first harvesting equipment (a grain binder); and built the Harvester Works to make its own binder.

Deere & Company had now assembled some of the best farm tools and their makers to become a major agricultural implement manufacturer. But it hadn't approached the hot topic of building gasoline traction engines—an area where its competitors were already "making smoke."

1929 Deere GP

This 1929 GP has a rear-mounted sickle-bar tractor mower. The built-in PTO powered the mower and Deere's Power Lift raised the sickle bar with engine power. Rubber tires mounted on factory rims modernized this tractor and extended its farm utility. The GP model series included the GPO orchard tractor, the GPP potato tractor, the GPWT wide tread, and the GPO Lindeman track conversion. Phyllis and Wayne Pokorny of Malmo, Nebraska, collected and restored this GP.

Deere had shied away from building steam engines and threshing machines and was apparently reluctant to enter the tractor field. The Deere board, controlled by Deere family members, was conservative with company money. They had seen several ambitious tractor companies launch and then quickly sink.

But pressures from its branch house and dealers to make a tractor were mounting, and in 1912 Deere authorized the development of a tractor plow. The resulting three-wheeler was similar to the Hackney Motor Plow, with an under-mounted three-bottom plow. Work on it stopped in 1914, and Joseph Dain Sr., who joined Deere at the acquisition of his company, was authorized to develop a tractor for Deere with a targeted market price of $700. His design was also a three-wheeler (a popular concept then), but had all-wheel drive. The first one was built in 1915, then tested, improved, and retested.

More test units were built and field tested in 1916 and 1917. In September 1917, Deere decided to build 100 of the Dain-designed all-wheel-drive tractors. Before that preliminary run was completed,

1933 Deere GPWT
The last of the GP tricycle-type row-crop tractors was the Model GPWT with over-the-top steering. It was produced until 1934. John Deere made several versions of wide-tread models beginning in 1929. This 1933 GPWT on factory rims for rubber tires has the tapered rear hood design that helped the operator see the crop rows being cultivated. John Deere advertisements claimed "new and improved four-row cultivating and planting" for the model. Well-known John Deere collector Verlan Heberer of Belleville, Illinois, restored this wide-tread tractor for his collection.

1934 Deere A

John Deere's most popular tractor of all time was its Model A row-crop tractor, introduced in 1934. The Model A was a second-generation row-crop machine with improved performance over Deere's first row-crop: the GP. Splined single-piece rear axles mounted with large-diameter rear wheels gave the A adequate under-axle crop clearance without the need for drop gearboxes. Rear tread could be easily adjusted by sliding the wheel hubs to the desired width. The Model A had a new optional hydraulic lift for cultivators and mounted implements. Bill Ruffner of Bellevue, Nebraska, posed his 1934 Model A under the state's largest cottonwood tree.

1935 Deere AR

The standard-tread version of the Model A was the Model AR, introduced in 1935. The engine in the Model A burned distillate at 975 rpm and could handle two 16-inch plows. The Model A was available in several different configurations. This 1935 Model AR is equipped with steel lugged wheels. By the late 1930s, most tractors were being delivered on rubber tires. Nebraska tests rated the Model A on steel wheels at 18.72 drawbar horsepower and 24.71 belt horsepower in 1934. Kent Kaster of Shelbyville, Indiana, owns this sturdy AR.

Deere & Company bought the Waterloo Gasoline Engine Company of Waterloo, Iowa, on March 14, 1918. Although the 100 Dain tractors were manufactured, the Waterloo Boy doomed the all-wheel-drive design. The final price of the Dain tractor had risen to $1,700 compared to only $850 for the simpler Waterloo Boy tractor.

FROELICH HERITAGE

The Waterloo Gas Engine Company began as the Waterloo Gasoline Traction Engine Company in 1893, formed by John Froelich and others. A year earlier, thresherman Froelich put together a gasoline tractor that he used successfully in his threshing season in South Dakota. Froelich had mounted an upright, 20-horsepower, Van Duzen single-cylinder gasoline engine on a Robinson steam engine chassis and then devised gearing for propulsion. Not only could his gasoline traction engine go forward and backward, but once belted to his threshing machine, it threshed 62,000 bushels of grain during the fifty-two day threshing run. Many historians consider the Froelich machine the first practical tractor. It was not, however, a commercial success.

Froehlich's new Waterloo, Iowa–based company built four more tractors in 1893; the two that sold were returned for major inadequacies. Although another tractor was built in 1896 and yet another in 1897, the young company decided to concentrate instead on gasoline engines and Froelich, whose primary interest was the tractor, dropped out of the firm in 1895.

1937 Deere AW
A wide-front 1937 Model AW pulls a John Deere 290 two-row high-speed check-row planter. French & Hecht wheels with their round spokes suggest this Model AW came from the factory on rubber. Tested on rubber tires in 1939, the Model A produced 26.2 drawbar horsepower and 29.59 belt horsepower, a decent gain of 7.48 drawbar horsepower due to its rubber tires. Jim Finnigan of Shirley, Illinois, owns this beautifully restored Model AW and planter.

By 1911, renewed interest in the tractor spurred more work on the machine, and by 1914 an early version of the Waterloo Boy Model R was built with an integral cylinder block for its horizontal two-cylinder engine. An opposed crankshaft set the pattern for later Deere engines. By the time Deere purchased the Waterloo Gasoline Engine Company, the Model N had been launched and more than 8,000 Waterloo Boys had been sold. The main difference between the Model R and the N was the N's two-speed transmission. The R had only one speed.

In 1920, when the University of Nebraska got its tractor tests underway in Lincoln, the Waterloo Boy Model N was the first tractor tested. The 6,183-pound machine lived up to its 12/25-horsepower rating, proving 12.1 drawbar horsepower and 25.51 horsepower on the belt. Its kerosene-burning engine had a 6.5x7-inch bore and stroke. Improved versions of the Model N continued in production until 1924. Among other changes, automotive-type front-wheel steering replaced the earlier fixed axle with chain-wrap steering. In all, about 20,000 Waterloo Boys were sold. Even though they were made and sold by John Deere, none of them carried that name.

Deere came late to the tractor business and was in a hurry to catch up. The competition, especially from the Fordson, was heating up. Sales of the lightweight 2,700-pound Fordson tractor were soaring. During the time Deere was getting on line with the Waterloo Boy in 1918 to 1920, Ford sold 158,000 Fordsons. By comparison, about 13,700 Waterloo Boys were sold in the same period. Times were tight at Deere in the economic recession after

1937 Deere A

An umbrella and a good breeze supplied the operator natural air conditioning while cultivating with this 1937 Model A. The hand-lift two-row cultivator took some muscle at row ends to raise it before and after the turn. Four-row cultivators made the Model A a very productive cultivating tractor. Phyllis and Wayne Pokorny of Malmo, Nebraska, collected and restored this Model A. Red paint on the cultivator shanks, a John Deere convention, adds color to this beautiful rig.

World War I, but if the firm was going to stay in the tractor business it needed an improved machine. Would it be a four-cylinder tractor? Economics of manufacture and maintenance prevailed, and Deere produced two-cylinder tractors for the next forty years.

DEERE WITH A D

Between 1919 and 1922, experimental versions of a new tractor based on the Waterloo tractor were built and tried. Finally, in 1923, the John Deere Model D 15/27-horsepower tractor came out. It was the first tractor to carry the John Deere name. The Model D was of frameless construction, had an all-enclosed final drive, pressure lubrication, and a standard wheel configuration. The D was also a three-plow tractor, designed to run on kerosene or distillates. Like other kerosene-fueled tractors of its day, it used water

from its cooling system to reduce pre-ignition or knocking when under heavy loads.

Deere's Model D was produced from 1924 to 1953—thirty years, the longest production run for one Deere model. During that time, some 160,000 Model Ds rolled off factory lines. In its 1924 Nebraska tests, the Model D produced 22 horsepower at the drawbar and 30 belt horsepower. By the end of production in 1953, the Model D's power had been upped to 38 on the drawbar and to 42 belt horsepower.

With the advent of IHC's Farmall in 1924, Deere couldn't ignore the need for a row-crop or general-purpose tractor. The Farmall was ransacking Fordson sales, and it became evident to John Deere's marketing group that a tractor, designed to cultivate row crops as well as pull tillage tools and run belt applications, was a real necessity for their tractor line.

1936 Deere A
Factory-supplied rubber tires equip this pristine 1936 John Deere Model A. On rubber tires, the A could handle three 14-inch bottom plows under most conditions. More than 328,400 Model As sold during its production run from 1934 to 1952. Most of the sales were of this tricycle-geared version. Tom Manning of Dallas Center, Iowa, found and restored this good-looking Model A.

THREE ROWS AND A LIFT

Work on a three-row cultivating tractor began in 1926, and following two years of prototype testing, production started in 1928. Originally designated the Model C, the new cultivating tractor was renamed the GP for "General Purpose," possibly to avoid confusion between the spoken "D" and "C" sounds. The GP had a standard four-wheel configuration. An arched front axle and drop gear housings on the rear axles gave it clearance for row-crop cultivation of three rows. The tractor straddled the center row with the front wheels, and its driving wheels centered between each outside row. It resembled the Model D in many respects, but was smaller and had more crop clearance.

The GP introduced, for the first time ever, a motor-driven power lift for its front-mounted cultivator and rear-mounted planter. With the lift's introduction, power farming now considered the comfort of the operator. But the GP was an early disappointment. Not all farmers were thrilled with the three-row concept, and some had troubles seeing the row when cultivating row crops.

Deere rushed to remedy these problems, and by the cropping season of 1929, the fix was available. It was the Model GPWT, or GP Wide Tread, a tricycle design with wide-set rear wheels and a front-mounted two-row cultivator—more like the popular Farmall. In 1931, the GP got a boost in horsepower from its original 10/20 level, moving it up to 15/24 horsepower. In 1932, the GPWT's hood was tapered at the back to aid visibility, and the steering was modified to an over-the-top configuration with improved steering characteristics.

1938 Deere AO
Deere engineers trimmed down a Model A in height and width and added sheet metal shrouds to form this early streamlined tractor. Their creation was the Model AO (often called AOS) orchard tractor. Its farm job was to slip under and through orchards and groves without damaging the valuable trees. Deere later streamlined and modernized its tractors with help from an industrial design specialist. This 1938 AO was collected and restored by Edwin L. Brenner of Kensington, Ohio.

1936 Deere B
A John Deere row-crop tractor two-thirds the weight and power of the Model A was developed and announced by John Deere in 1935: the new Model B. Its distillate engine had a 4.25x5.25 bore and stroke and revolved at 1,150 rpm to make its Nebraska-measured 11.84 drawbar horsepower and 16.01 belt horsepower. Don McKinley and Marvin Huber of Quincy, Illinois, located and restored this 1936 Model B, shown with a two-row hand-lift BB 221 cultivator.

The later-version GPWT added an adjustable cushion spring seat, spark and throttle levers mounted on the steering wheel support, and individual rear wheel brakes for shorter turns between row-pairs when planting or cultivating. Various versions of the GP were produced, including an orchard version GPO, a potato version or P Series, and some crawler-tracked versions put together by Lindeman Power Equipment Company of Yakima, Washington.

The GP stayed in the John Deere line until 1935, and the GPWT was phased out in 1933.

A POPULAR DEERE DUO

Deere's second-generation row-crop tractor was its all-time best seller. Building upon early GP experiences, Deere introduced the Model A in 1934. It had a tricycle-gear row-crop design with the

Deere Model B
Deere's Model B tractors were designed with one main purpose in mind: to replace the last teams of horses on American farms. Here a Model B pulling a corn picker shares the field with a horse-drawn wagon, proving that horses still had a place as field workers on many farms well into the 1940s.

now familiar John Deere two-lunger powerplant. But under the green and yellow paint, it was all new. It had infinitely adjustable rear-wheel tread on splined single-piece rear axles, a one-piece transmission case, a hydraulic power lift that both raised and "cushion-dropped" implements, a centerline draft, and differential brakes geared directly to the large drive gears. All of these features made the Model A a useful, very efficient row-crop tractor with easy handling.

The Model A was rated as a two 16-inch-plow tractor, and its 1934 Nebraska tests measured 18.72 horsepower on the drawbar and 24.71 horsepower on the belt from a 5.5x6.5-inch distillate engine turning at 975 rpm. Production figures indicate more than

1936 Deere B

Lighter farm jobs, like pulling this 10-foot 1935 John Deere tractor binder, were suggested as uses for the Model B. Unlike earlier binders that were powered by a ground drive through the binder bull wheel, the tractor binder got its power from the tractor's PTO. A bundle accumulator on the right side of the binder dumped bound bundles together, so they could later be stacked into weather-shedding shocks to await later threshing. Don McKinley and Marvin Huber of Quincy, Illinois, restored these gems from the past.

328,400 Model As, in various configurations, were made from 1934 to 1952. Most were the row-crop versions.

The Deere machine aimed at replacing the last team on the farm was the Model B. It too had a long and successful production run. More than 322,200 Model Bs came off factory lines between 1935 and 1952. Advertised as being two-thirds the size of the A, the little B made 11.84 horsepower on the drawbar and 16.01 on the belt when first tested at Nebraska in November 1934. Its 4.25x5.25-inch engine turned at 1,150 rpm. The tractor's shipping weight was 3,275 pounds, compared with 4,059 for its bigger brother, the Model A.

Many specialized versions of the A and B were produced, including the standard-tread AR and BR and the orchard-equipped

1935 Deere Grain Binder
The grain binder was a much-needed labor-saving implement, eliminating the need for small grains to be harvested by hand with scythes. With the binder, the whole plant was cut, bundled, shocked, and later hauled to the threshing machine. Farm tractors soon replaced most of the horses once used in harvesting.

AO and BO. Wide-row versions were designated AW, narrow-row as AN, and high-crop clearance models as ANH and AWH. The Model B was also available in similar specialized versions. The Model As and Bs were the first John Deere tractors available with pneumatic rubber tires in the early 1930s. In 1936, the orchard version of the A was streamlined to become the AO Streamlined, or AOS as today's collectors call it. More streamlining was on the way.

FEWER HORSES BUT MORE HORSEPOWER

Individual farm size increased as mechanization improved the productivity and efficiency of farmers. One man could do more work than ever before and if given a larger tractor, he could do

even more. The tractor folks at Deere saw the need for a larger row-crop tractor for the larger farm market. In 1937, Deere responded with the three-plow Model G. It resembled the A and B, but had a stouter 6.12x7-inch bore-and-stroke engine that produced 20.7 drawbar horsepower and 31.44 belt horsepower at 975 rpm. It was heavier too, at 6,150 pounds. It could pull a 10-foot tandem disk or easily handle four-row planters and cultivators. Available only in row-crop configurations, the G Series was in production from 1938 to 1953, and more than 60,000 were sold. Over its production run, the G's power and weight increased, and a 1947 Nebraska test showed the G put out 34.49 drawbar horsepower and 38.10 belt horsepower when equipped with rubber tires.

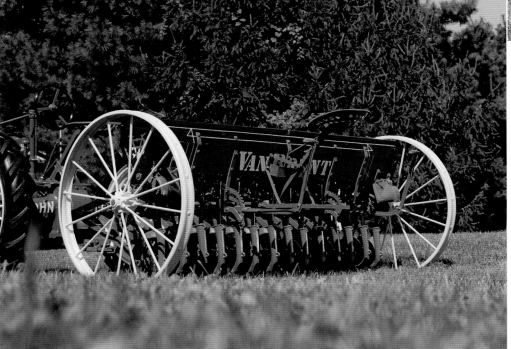

1935 Deere B with Grain Drill

The drilling of small grains with a grain drill was another operation that could be handled by the smaller Model B. This 1935 Model B tractor on factory rubber tires is hitched to a 1930s John Deere Van Brunt grain drill. It was used for planting small grains like wheat, oats, and barley, which were "drilled" in rows only 7 inches apart. The pan seat behind the drill's seed box is a leftover from days when the drill was pulled by a team of horses. To adapt the grain drill to tractor use, its long wooden horse tongue was cut down and a new tractor hitch installed.

This tractor was restored by Kent Shriver and the grain drill by Don McKinley and Marvin Huber, all of Quincy, Illinois.

In 1937 Deere introduced another tractor—this one on the lower end of the power scale. The Model 62 utility tractor was a vast departure from the other Deere models of the day. It was powered with a Deere-designed, Hercules-built, vertical two-cylinder engine. Its output was 7.01 horsepower on the drawbar and 9.27 on the belt. The 3.25x4-inch engine revved at 1,550 rpm. The Model 62 was built as a one-plow, one-row unit for small operators, including "truck farm" vegetable growers.

Innovative firsts on the series included the powertrain offset to the left with the operator's position offset to the right. The design made for better forward and under-tractor visibility while carefully cultivating rowed plants. A foot-operated clutch replaced the Deere's hand clutch. A belt pulley was available, but no PTO. Seventy-nine Model 62s were built in 1937, before the tractor was designated the Model L.

Styling or streamlining the L Series came in 1939. Later versions included the LA, with power increased by 4 horsepower and

1935 Deere B
Kay Brunner cast-steel wheels from California are featured on this 1935 Model B, rather than the spoked wheels furnished by the John Deere factory. This 1935 model also still carries the "General Purpose" designation on its hood. By 1947, the little Model B was showing horsepower tests of 24.62 drawbar and 27.58 belt when burning gasoline and moving on rubber. Ken Smith of Marion, Ohio, collected and restored this Model B.

1939 Deere BR

Introduced in 1935, the Model BR soldiered on until 1947 as the standard-tread model of the B. Robert E. Waits of Rushville, Indiana, rebuilt and restored this 1939 BR. Waits' research shows the little tractor was shipped new to Yorkton, Saskatchewan.

24-inch rear wheels (up two inches from the prior 22), and the LI industrial version painted highway yellow. Electric starting and lights were available in 1939.

LINES OF THE TIMES

By the 1930s, more and more industrial products were being redesigned to add aesthetically pleasing lines to their sometimes knobby utilitarian features. In 1937, Deere hired noted industrial designer Henry Dreyfuss and his group to help "style" its Model A and B tractors. Dreyfuss' streamlined Model A and B tractors were introduced in 1938.

Replete with new styling, a new small row-crop, the Model H, came out in 1939. It was a handy two-row tricycle resembling the A and B, but was smaller with a 3.56x5-inch bore-and-stroke two-cylinder engine turning over at 1,400 rpm. The belt pulley on the H ran off the camshaft rather than the crankshaft, as on Deere's larger tractors; this resulted in a higher pulley location and a reversed rotation compared with the other tractors. The Model H's 1938 Nebraska tests showed it produced 12.48 drawbar horsepower and 14.84 belt horsepower. Its production run was from 1939 to 1947, with more than 60,000 coming off factory lines. High-clearance versions, the HWH and HNH, were also available.

The rugged Model D finally got its styling facelift in 1939, and the Model G was updated to the six-speed Model GM in 1943. By then, rubber tires were standard. Only rubber shortages during World War II caused a temporary return to steel wheels. Tractor model introductions were nearly curtailed at John Deere during the war years, as the factories and engineering departments concentrated on defense work.

Deere BW-40
The BW-40 was designed for use in narrow-spaced crops. It used special narrow front axles and a rear axle case from the orchard version of the B to allow an extra narrow spacing of the rear wheels. Bruce D. Aldo of Westfield, Massachusetts, has this one in his collection of rare Deeres.

1937 Deere BI
The Model BI was the industrial version of the B. Yellow paint made the BI more visible on the highway. Kent Kaster of Shelbyville, Indiana, owns this 1937 machine, which has the inverted mufflers and padded seat used on Deere's industrial machines.

1945 Deere BO Lindeman Crawler
Before John Deere made its own tracklayer tractor, it furnished BO chassis without wheels to Lindeman Manufacturing, Inc., of Yakima, Washington. Lindeman installed its crawler undercarriage, and the John Deere Lindeman Model BO crawler was the result. Fruit growers in Washington State liked the low-profile machine for its ability to work on the steep slopes in the orchards there. Don Rogers of Atlanta, Illinois, found and restored this 1945 Lindeman BO crawler. Deere bought the Lindeman company in 1945 and based its later crawler tractors on the Lindeman design.

CREATURE COMFORTS BECOME STANDARD

Postwar, Deere began making electric starting and lights standard equipment on its tractors. Also available after 1947 was Powr-Trol, allowing hydraulic control of pulled implements. Roll-O-Matic became available on tricycle fronts to smooth out some of the front-end bumps.

Padded operator seats were welcome additions on the Model M tractors introduced in 1947. The new Dubuque, Iowa–built machine came out as a standard-tread general-purpose utility tractor.

It was the first in a line of vertical two-cylinder tractors. The M's 4x4-inch engine operated at 1,650 rpm, making it one of the fastest turning engines Deere had made to that time.

The M was joined by the tricycle-version row-crop Model MT in 1949. Both the M and the MT had Touch-O-Matic hydraulic controls for integral-mounted implements. The MT's dual system could be set to raise and lower front or rear implements separately, at the same time, or on the right and left sides independently. Horsepower output for the M and MT tractors was just over 14

horsepower at the drawbar and 18 on the belt. The M was built from 1947 to 1952 and the MT from 1949 to 1952. More than 40,500 Model Ms and 25,800 MTs were made.

Brother to the M and MT was the MC, the first Deere-designed and Deere-made crawler. Deere had long-supplied Model BO chassis to the Lindeman factory in Yakima, Washington. Lindeman then mounted its crawler units to the chassis, making tracklayers for orchard use in the area. In 1946, Deere acquired Lindeman and, in 1949, built its own track-type

tractor based on the same vertical two-cylinder engine used in the M and MT tractors.

The MC crawler and the industrial version of the M, the MI, became the basis for Deere's industrial division. In 1950 Nebraska tests, the new Deere crawler produced 18.3 drawbar horsepower and 22.2 belt horsepower. A credit to the MC's track design was its 4,226 pounds of maximum pull, compared with only 2,329 pounds for the rubber-tired Model M. The MC came on 12-inch-wide tracks, but 10-inch or 14-inch widths were available. The

1938 Deere G
"The new John Deere Model G, for the large row-crop farm," proclaimed advertisements for the new tractor announced in 1938. The larger tractor answered farmer interest in a more powerful row-crop machine. The Model G was a full three-plow tractor with a heftier 6.125x7-inch bore-and-stroke engine. Wayne Bourgeois of Kahoka, Missouri, collected this early "low radiator" version. Later production featured a larger radiator that was installed to cure an engine overheating problem. The Model G tested in 1937 on steel wheels at Nebraska, on distillate fuel, turned in 27.63 drawbar horsepower and 35.91 belt horsepower.

Opposite: **1938 Deere B**
The frame on the Model B was extended 5 inches in 1937 by John Deere to make mid-mounted implements interchangeable between the Model A and B tractors. This 1938 Model B was made just before the John Deere models were given their new styled look. Nathan Anderson of Stratford, Iowa, rebuilt and refinished this tractor after it had spent its working lifetime on his father Eldon Anderson's farm.

tread widths available ranged from 36 to 42 inches. About 6,300 of the MCs were made between 1949 and 1952.

DEERE'S DIESEL

Deere's first diesel tractor, a powerful replacement for the venerable Model D, came out in 1949. Based on engineering and design work started in the war years, the Model R was Deere's most powerful tractor to date. Its 5.75x8-inch diesel operated at 1,000 rpm and gave it 45.7 drawbar horsepower and 51 belt horsepower in its

Nebraska test. A small electric-start two-cylinder gasoline-starting motor put the R's diesel into motion. The big tractor had a shipping weight of almost 10,400 pounds and showed a maximum pull of 6,600 pounds in Nebraska tests. More than 21,000 Model Rs were produced between 1949 and 1954.

Introduction of new models in all sizes accelerated after the war, especially in the early 1950s when tractor production caught up with demand, and tractor manufacturing and sales again became a competitive business. More horsepower and more features became

1945 Deere GM
A new version of the John Deere Model G received a new designation—GM—during World War II. The GM featured a six-speed transmission, the new Henry Dreyfuss styling, and optional electric lights and starter. After the war, the GM's designation became the G again. A 1947 Model G was tested at Nebraska on rubber tires and fueled with gasoline. It produced 34.49 drawbar horsepower and 38.10 belt horsepower, a substantial improvement over the previous test. Retired Deere dealer Bill Zegers of Newton, Iowa, found and restored this 1945 Model GM.

Above: 1937 Deere 62
A really small tractor for the commercial market gardener was introduced in 1937. The Model 62 was made in small numbers until it became the unstyled Model L. The tractor's design offset the engine and drivetrain to the left with the operator's seat installed on the right for improved visibility and control of the one-row cultivator. The tractor's engine was a two-cylinder upright design that delivered 7 drawbar horsepower and 9 belt horsepower. Ronald Jungmeyer of Russellville, Missouri, owns this collectible Model 62.

Left: 1938 Deere LI
This 1938 Model LI, a Deere industrial tractor, was equipped with a sickle bar mower for highway right-of-way maintenance. This one is owned by Ron Jungmeyer of Russellville, Missouri.

1941 Deere LA

The most powerful of the small John Deere tractors was the Model LA. This 1941 version has a 14.34-horsepower engine. Solid frame rails and cast-iron 24-inch wheels boosted the LA's operating weight to 2,900 pounds. The mighty little LA has become a favorite of collectors, as it handily fits in the bed of a pickup truck. Phil Graber of Mount Pleasant, Iowa, owns this Model LA equipped with lights and a starter. The Model L and LA tractors were made in Moline, Illinois, at the John Deere Wagon Works.

Wartime John Deere advertisement
During World War II, tractor production nearly came to a standstill as manufacturers like John Deere put their efforts into producing military machines and even artillery. Yet Deere didn't forget about the farmers at home, lauding women's effort in keeping the fields tended with this ad that says "the hand that rocked the cradle is the extra hand that's fighting for democracy."

1941 Deere LA
A good view of the LA tractor's working area is provided by the offset engine and drivetrain.

the name of the game. In the ten years from 1949 to 1959, Deere introduced six major model series.

The first of John Deere's numbered tractor series was introduced in 1952. The Models 50 and 60 were worthy replacements for Deere's B and A models. New sheet metal styling with small vertical grooves in the radiator screens gave the series a new look. Duplex carburetion with a carburetor for each cylinder helped increase performance. The power take-off operated independently

of the transmission or transmission clutch to provide continuous power for drawn implements. Hydraulics were also improved. A live high-pressure Powr-Trol allowed for faster lift of heavier equipment, and it operated independently of clutch or PTO. In 1954, the series introduced a couple of industry firsts: factory-installed (but optional) power steering and rack-and-pinion adjustment of rear-wheel tread.

The Model 70 joined the numbered series in 1954 as the replacement for the G. It was offered in gasoline, "all fuel," or LP versions, as were the 50 and 60 models. In 1954, the Model 70 was available with a diesel engine as Deere's first diesel row-crop tractor. Drawbar and brake horsepower tests at Nebraska resulted in 20.62 and 26.32 horsepower for the Model 50, 27.71 and 35.33

1939 Deere BNH
New York industrial designer Henry Dreyfuss gave the John Deere Models A and B a new stylish look in 1937. Eventually, the entire John Deere tractor line showed the effects of his design work, which took some of the rough edges off of the machines. With a single front wheel, 40-inch rear wheels, and long axles, the BNH pictured here could be set up for high- and wide-clearance applications in vegetable production. Tom Jarrell of Wilmington, Delaware, owns this 1939 styled Model BNH.

horsepower for the Model 60, 42.24 and 48.29 for the Model 70 gasoline, and 34.25 and 43.77 for the diesel Model 70. Several different front- and rear-axle versions were available on the numbered series, as were high-crop models, for those growers needing high-clearance capabilities. The Models 60 and 70 were also made as standard-tread models, and the 60 was available as an orchard tractor. More than 33,000 Model 50s were made from 1952 to 1956, 64,000 of the various 60 models, and 43,000 of the Model 70 were manufactured from 1953 to 1956.

Upgraded versions of the Dubuque-made Model M were introduced in 1953 as numbered models. They were the Model 40 standard, the Model 40 tricycle, and the Model 40 crawler. They also wore the new grille treatment of the 50, 60, and 70 models. Improved hydraulics and other systems were similar to those on the larger tractors. When compared to the earlier Model Ms, the Model 40s had 15 percent more horsepower. The new Model 40 crawler sported four or five track rollers, depending on track length. The 40 Series was produced between 1953 and 1955.

1945 Deere H
A new smaller, lower-priced John Deere tractor "that handles every job on the small farm and many jobs on the large farm" was the way the Model H was described at its 1939 introduction. The H was released as a styled tractor. It put out 11.67 drawbar horsepower and 14.84 belt horsepower in pre-production testing at Nebraska. Options available included a single front wheel, fenders, a radiator shutter instead of a curtain, independent hydraulics, and electric starting and lighting. Don Ward of Chula, Missouri, collected and restored this immaculate 1945 Model H with electric lights and starter.

1948 Deere M

Envisioned, designed, and built to compete with the popular Ford-Ferguson 9N tractors, the John Deere Model M finally arrived in 1947. Built in a new factory in Dubuque, Iowa, the Model M was also designed to replace the small LA and LI models built in Moline and the Model H, BR, and BO built in the Waterloo factory. The Model M had a new upright two-cylinder 4x4-inch engine that ran at 1,650 rpm and featured Touch-O-Matic hydraulic implement control. Nebraska tests in 1947 showed the M produced 18.15 drawbar horsepower and 20.45 belt horsepower when fueled with gasoline. Bill Zegers of Newton, Iowa, collected this 1948 Model M that is equipped with a John Deere blade.

Meeting the demand for more horsepower in its largest tractor, Deere replaced the Model R with the diesel Model 80 in 1955. The big 80 produced a whopping 46.32 horsepower on the drawbar and 57.49 on the belt. It was also upgraded to include the same hydraulic systems and live PTO as the other numbered models had. The Model 80 was produced from 1955 to 1956 when it was upgraded to a new model designation.

The two-tone 20 Series arrived in 1956. In addition to yellow on the wheels, the 20 Series had a yellow stripe on the bottom of the hood continuing down the sides of the radiator shroud. The Waterloo-made tractors—the 520, 620, and 720 models—also had new engines, featuring improved cylinder heads and pistons to increase combustion-chamber turbulence for more output and fuel efficiency. The Dubuque tractors included a smaller version,

1949 Deere MT

The tricycle-geared version of the M was the John Deere Model MT row-crop machine that went into production at Dubuque near the end of 1948. The MT engine and drivetrain were offset to the left to provide better vision on the right side of the tractor. The MT also featured the Quik-Tatch implement attachment. Quincy, Illinois, collector Ted Shriver's 1949 row-crop machine, pictured here, is equipped with Dual Touch-O-Matic with a split rockshaft to handle left- or right- or front- and rear-mounted cultivator gangs.

1949 Deere B and 1950 Deere A

The styled Model A and B tractors received some important updates in 1947. These included using pressed steel frames, making electric lights and starting standard equipment, putting batteries under the tractor's cushioned seat, adding modern steering columns, and offering gasoline or all-fuel engines. The gas engine in the Model A produced 34.14 drawbar horsepower and 38.02 belt horsepower, making it a full three-plow tractor. The Model B engine (at right) was upgraded by an increase in the engine bore from 4.5 inches to 4.6875 inches, and the engine was speeded up to 1,250 rpm. That gave the gas version Model B tractor 24.62 drawbar horsepower and 27.58 belt horsepower. These updated As and Bs eventually became known as "electric" models.

the Model 320, rated as a one- to two-plow tractor with the hydraulics of the other models. The Model 420 superceded the 40 Series and was made in tricycle, standard, high-crop, utility, wide-tread, low-profile, and crawler versions. The 820 diesel replaced the 80 as the big gun in the line. After a factory power tweak, it pulled 52.5 drawbar horsepower and 64.26 belt horsepower in its 1957 Nebraska test.

THE LAST OF THE JOHNNY POPPERS

The world couldn't know, but the 30 Series introduced in 1958 would be the last of Deere's two-cylinder tractors. Operator com-

fort and ease of use were the hallmarks of the series. A new deep-cushioned seat positioned the operator between two new flat-top fenders with built-in dual lights. A new dash design tilted the engine instruments for a better view. The steering wheel column emerged from the dash with the wheel positioned on an angle for more comfortable steering. The Dubuque tractors, still using vertical two-cylinder engines, were the 330 and the 430 models, and they came in many tread and axle configurations. Dubuque also produced the 430 crawler.

From the Waterloo tractor plant came the 530, 630, 730, and 830 models and their derivatives. The 730 diesel had electric starting,

1952 Deere AW
John Finnigan of Shirley, Illinois, is the proud owner of this 1952 Model AW, shown pulling a John Deere 730 lister planter (which helps plant seeds deep in the soil in dry farming conditions). John restored this tractor as an FFA project in 1998 and was an award winner at the 1998 National FFA tractor restoration contest.

1951 Deere R
The first John Deere diesel tractor was the Model R, introduced in 1949. The R was developed during World War II. It put Deere tractors squarely in the horsepower race with 45.7 drawbar horsepower and 51 belt horsepower from 5.75x8-inch bore-and-stroke cylinders running at 1,000 rpm. The R was the largest and most powerful John Deere tractor to date. Chad Reeter of Trenton, Missouri, restored this Model R—equipped with deep-tread rice tires—with help from friend Mel Humphreys.

1953 Deere AR
This 1953 John Deere Model AR tractor was manufactured for export to Canada. It later returned to the United States as a collector-owned machine. The AR was the last of the Waterloo-built Deere tractors to be styled. The AR was introduced in 1949 with a six-speed transmission with an extra-low "creeper" gear of 1.3 miles per hour. The AR, or standard tread, was a three-plow tractor that produced 34.9 drawbar horsepower. This tractor is in the extensive John Deere tractor collection of Jim and Doris Finnigan of Shirley, Illinois.

eliminating the gasoline-starting engine. The very last two-cylinder Deere model introduced was the Model 435 diesel, powered by a General Motors two-cycle diesel engine. The 435 diesel was the first Deere tractor tested at Nebraska that received a PTO horsepower rating rather than belt horsepower rating.

A sign of things to come coincided with the surprise introduction of a monster four-wheel-drive Deere tractor in the fall of 1959. The Model 8010 had a company rating of 150 drawbar horsepower from a 215-horsepower engine. It could pull a mounted eight-bottom plow at seven miles per hour, lift the plow at the row end, and turn sharply back into the next furrow. Articulated steering, power steering, and air brakes aided in its control. Another version, the 8020, came out in 1960 to add an eight-speed Syncro-Range transmission, an oil-cooled clutch, and other improvements. Somewhat ahead of their time, neither version sold well.

NEW GENERATION OF POWER

The world learned on August 30, 1959, what the Deere engineers had been doing for the past six years. That date marked the unveiling to Deere dealers of the New Generation of Power tractors in Dallas, Texas. The new models—the 1010, 2010, 3010, and 4010—were all new from front to back and top to bottom. Proven Deere concepts, such as weight concentrated on the rear wheels, were achieved by placing the engines to the rear, close to the transmission and differential. Up front were the large-capacity fuel tank and radiator. And powering the new machines were all-new vertical four- and six-cylinder engines. There was not one two-cylinder engine in the line.

The four-cylinder 3010 and six-cylinder 4010 were available as diesels, as well as gasoline or LP-fueled versions. High horsepower-to-weight ratios gave the new tractors the advantage of operating at higher speeds while using less power to move; that added efficiency to their obvious advantages of increased power.

1954 Deere 60
Farmer-collector-restorer Bob Hild of Webster City, Iowa, bought this Model 60 new in October 1954, after he returned home from U. S. Army duty in South Korea. The new numbered series replaced the models A, B, G, M, and R with the models 60, 50, 70, 40, and 80. The tractors used a cast-steel frame, instead of the pressed-steel frame of the previous Model As and Bs. The Model 60 produced 36.9 drawbar horsepower with gasoline in Nebraska trials.

1953 Deere 50

John Deere's venerable Model B was replaced in 1952 with the new Model 50. Part of the new numbered series, the Model 50 came off factory lines with duplex carburetion, live PTO, high-pressure Powr-Trol hydraulics, and rack-and-pinion rear-wheel adjustment. The 50's styling was new too, with a vertically fluted perforated metal forming the radiator grille. New sheet metal also marked the tractor hood. The gas Model 50 tested 27.5 drawbar horsepower at Nebraska. Bob Hild of Webster City, Iowa, owns this 1953 tractor.

1954 Deere 60

Wide front axles, adjustable for row-crop use, were a popular option on the John Deere Model 60. In addition to duplex carburetion, the 60 offered an all-weather manifold, independent PTO, optional power steering, and continuous high-pressure Powr-Trol hydraulics. Operator control was augmented with a longer hand clutch and throttle controls that were easier to reach. Douglas Latham of Earlham, Iowa, owns this 1954 tractor that worked on his father Earl Latham's farm near Murray, Iowa.

1954 Deere 60S LPG
This 1954 John Deere Model 60S was equipped to burn liquefied petroleum gas (LPG) and came with rice tires for extra grip in soft southern fields. It has the high seat position that was used on the later standard-tread models and was derived from the popular wide-front row-crop models. Ford Baldwin of Lonoke, Arkansas, owns this tractor.

The 4010 produced 73.65 drawbar horsepower, but weighed less than 7,000 pounds. The 3010 diesel pulled more than 50 horsepower at the drawbar and was a full four-plow tractor.

Added to the other firsts, the new series of tractors had closed-center hydraulics for live implement-raising muscle in three circuits. Power steering and power braking were also served by the new variable-displacement pump working from a common reservoir of hydraulic fluid. New eight-speed Syncro-Range transmissions matched power and speed to job demands.

Other power sizes in the series were the 1010 with 30 horsepower available at the drawbar and the 2010 with up to 40 drawbar horsepower. The 1010 had a utility configuration, but with adjustable front and rear tread to handle row-crop work. The 2010 was offered in tricycle row-crop or row-crop utility versions.

The years following the introduction of the New Generation tractors proved the wisdom of Deere's gamble. Farmers widely accepted them, and Deere's share of the U.S. wheeled tractor

1955 Deere 60 High-Crop
Working in tall-growing crops was the specialty of this 1955 Model 60 High-Crop tractor. A crop clearance of 32 inches under the front and rear axles allowed this tractor to work in many applications. Built from 1952 through 1956, the Model 60 was available in different versions to burn LP gas, gasoline, or all-fuel. Jim and Doris Finnigan of Shirley, Illinois, own this tall rig with its factory power steering.

1956 Deere 70 Diesel
John Deere's first row-crop diesel tractor was the Model 70. The 70 was also available in gasoline and LP versions. The 70 was a worthy successor to the old Model G, which dated back to 1937. The 70 diesel's two-cylinder 6.125x6.375-inch engine showed 45.7 drawbar horsepower and 51.5 belt horsepower when tested at Nebraska in 1953. It also set a new fuel economy record. Factory power steering helped the operator handle the 9,028-pound machine. An auxiliary gas V-4 pony engine fired up the big diesel. Owner Ed Hermiller of Cloverdale, Ohio, added the shiny muffler to this 1956 Model 70 diesel to replace the original one.

1953 Deere 70 LPG
Proof of the Model 70's power is the mounted No. 813 John Deere plow. The three 16-inch plow bottoms are the right load for the tractor. Front frame weights help counterbalance the rear-mounted plow. Wheel weights add traction to the tractor drive wheels. The four-spoke steering wheel indicates that this tractor lacks power steering. Bob and Mark Hild of Webster City, Iowa, restored and own this 1953 John Deere Model 70 LPG.

market shot up from 23 percent in 1959 to 34 percent by 1964. In 1963, Deere outpaced rival International Harvester Company to become the sales leader in U.S. farm and light industrial equipment.

LEADING THE HORSEPOWER RACE

The first tractor to supply more than 100 horsepower to its two rear wheels was the 1962 John Deere diesel standard-tread Model 5010. By 1965, its successor, the 5020, pumped out 115.73 drawbar horsepower and 133.25 PTO horsepower in its Nebraska tests. Those ratings were furnished by the 5020's John Deere 531-cubic-inch six-cylinder diesel. Horsepower was soaring! Deere offered its big tractor in a row-crop version in 1967. It was capable of cultivating twelve rows of 30-inch spacing.

Introduced in 1964, Deere's 3020 and 4020 models carried more horsepower than the New Generation line and were destined to become the company's modern classics. The 91.70-PTO

1953 Deere 40

With a design harking back to World War II and the M and MT models, the Model 40 was the Dubuque factory's contribution to the new numbered series. The tricycle-geared Model 40 was a popular two-row row-crop tractor. It had an improved three-point hitch hydraulic system that provided live hydraulic power when the engine was running. This 1953 tractor provided 17.6 drawbar horsepower to its farm tasks. Mark Hild of Webster City, Iowa, is the owner of this 40 with a comfortable cushioned seat.

1953 Deere 40N

This 1953 Model 40N has a single front wheel to adapt to narrow rows. The Model 40 was also available as the standard-tread 40S or the track-type 40C. Jeff Underwood of Dahlonega, Georgia, collected and restored this handy-sized tractor.

1955 Deere 40V

Only 328 of the Vegetable Special Model 40Vs were made between 1954 and 1955, making this tractor a rare collectible. The 40V's under-axle crop clearance was 26.5 inches. A four-speed transmission, three-point hitch, and belt pulley helped it do a variety of farm jobs. Ken Peterman of Webster City, Iowa, collected this 1955 model.

1955 Deere 40 High-Crop
Specialty crop growers who needed high-clearance features chose the Model 40 High-Crop tractor at their John Deere dealer. Crops up to 32 inches high could pass under the tractor without touching the axles. The 40 Series tractors had their working horsepower boosted to 22.9 drawbar horsepower with an increase in the engine operating speed to 1,850 rpm. This 1955 Model 40 High-Crop is owned by Joe Necessary of Heyworth, Illinois.

horsepower 4020 stayed in the line through 1972 and sold an impressive 177,000 units, making it the most widely sold John Deere model of its era. The sales of the Model 3020, a 65.28-PTO horsepower machine, for the same period totaled an impressive 86,000 tractors. Both models offered the popular Power Shift transmission, which allowed for clutch-less, on-the-go shifting.

The Model 7020, launched in 1970, was an articulated steering four-wheel-drive of 146.17 PTO horsepower with a Roll-Gard cab

placed to the rear of the front tractor section. The cab was above the pivot point of the two-piece chassis. The 7020 was joined in 1972 by a stronger Model 7520 with a turbocharged-intercooled 532-cubic-inch diesel operating at 175.82 PTO horsepower. Both models offered row-crop capabilities.

In 1971, John Deere offered not only turbocharging on its Model 4620, but it enhanced tractor performance to 135.62 PTO horsepower with intercooling. It was the first time in the industry

1956 Deere 320 Standard
A 20-percent power increase was featured in the new John Deere 20 Series models that began arriving in 1955. The two-tone models with extra splashes of yellow on the hoods and radiator shrouds were marked as new and different. Smallest was this 1956 Model 320 standard. It didn't get the extra power awarded the other models, basically staying a Model 40 at heart. Its engine produced at 22.4 drawbar horsepower. Dawn Finnigan of Shirley, Illinois, is the owner.

that an intercooler was used on a turbocharged tractor. The intake manifold intercooler cooled the engine's turbocharged air to supply more oxygen and power potential to the engine.

Throughout the 1970s, 1980s, and 1990s, tractor engine horsepower continued to reach higher and higher levels. By the dawn of the twenty-first century, tractors boasted 500 horsepower levels. Also by 2000, many longtime tractor manufacturers had merged with other companies—most falling victim to an agricultural recession in the 1980s. Among them was International Harvester Company, which became part of J. I. Case and Company in 1984.

Deere & Company survived that turmoil, and today is the world's leading producer of farm and forestry equipment. It also is a major maker of construction equipment in North America, as well as commercial and consumer lawn and turf care equipment and off-highway diesel power systems.

John Deere continues its dominance in farm equipment manufacture and sales. In 2005, Deere & Company produced total revenues of $21.930 billion. In second place in farm equipment sales is CNH Global NV, the holding company that includes the Case-IH line of farm equipment. CNH had 2005 revenues of $12.575 billion.

Above and right: 1958 Deere 420 High-Crop
This 1958 Model 420 High-Crop tractor had the tall legs needed to work in staked vegetables and other tall crops. The 420 models all shared the same vertical two-cylinder 4.25x4-inch bore-and-stroke engine, which turned at 1,850 rpm to produce 37.08 drawbar horsepower in the gas version. Early Model 320s and 420s came in all-green paint until they were "two-toned" in 1957. Jim Finnigan of Shirley, Illinois, collected this rather rare 420 High-Crop.

1956 Deere 420 Utility
John Deere's Highway Orange paint identifies this 1956 Model 420 Utility as an industrial version. The utility tractor's lowered profile aided its stability while maintaining highway right-of-ways. The mid-mounted sickle-bar mower and optional five-speed transmission (coupled to a two-position foot clutch) made this a handy mowing machine. The first position of the clutch allowed the mower to continue to run while the tractor stopped. Collector Paul Watral of Hauppauge, New York, found this tractor in Iowa.

1957 Deere 420 Utility LP
This 1957 John Deere Model 420 Utility LP is a delight for rare tractor collectors. Only six of them were made. This low-profile utility version has an optional direction reverser for quick changes of direction. The LP, or liquefied petroleum fuel option, is also rare on this model. John Deere offered nearly a dozen variations of the 420 model. Collector Jimmy Strube of Garden City, Texas, had this tractor restored by Paul Lehman of Perry, Iowa.

1958 Deere 420 LPG Crawler
Liquefied petroleum gas fueled this 1958 John Deere Model 420 crawler. It is painted the green and yellow of the agricultural crawler, but this tractor has the five-roller undercarriage more commonly used on the industrial model. This 420 crawler is equipped with a three-point hitch mounting a single shank-deep subsoiler. John Deere's tracklayer machines trace back to Lindeman Manufacturing of Yakima, Washington. Lindeman put its crawler undercarriage on BO chassis, creating the John Deere BO Lindeman tractor. Randy Griffin of Letts, Iowa, owns this Model 420 crawler.

Upper left: 1957 Deere 520

Replacing the Model 50 in the 20 Series line was the new John Deere Model 520. In 1958, the 520 included a new high-output engine and a new Custom Powr-Trol. The Model 520 was only available as a row-crop machine. Its extra power came from a boost in the compression ratio and an increase in engine speed—to 1,325 rpm. As a result, the 520's drawbar horsepower increased 25 percent to 34.17. The owner of this 1957 tractor is Don Rimathe of Huxley, Iowa.

Above: 1958 Deere 520V

A large single front tire and big 42-inch rear wheels on long rear axles made this 1958 Model 520 a vegetable cultivation specialist. The 520V could be configured to work in multiple narrow row crops. The large rear tires also gave it extra clearance for working in high staked crops. The 520, with a lineage tracing back to the Model B, had grown in size and power to rate as a three-plow tractor. Brent Liebert of Glidden, Iowa, owns this 520V.

Left: 1958 Deere 620 Standard

More power, better hydraulics, and more operator comfort came with the added yellow trim on the 620 model series beginning in 1956. This 1958 John Deere Model 620 Standard showed 44.16 drawbar horsepower, 20 percent more than before, in its Nebraska tests and came up to four-plow performance. The Model 620 was available in standard-tread, high-crop, orchard, and several row-crop versions. Deere engineers got more horsepower from the 60 engine when they shortened its two-cylinder crank throw from 6.75 down to 6.375 inches and increased engine speed to 1,125 from 975 rpm. Doris Finnigan of Shirley, Illinois, claims this "muscular" Model 620 Standard.

1958 Deere 620 Orchard LPG

Orchard tractors, like this 620 LPG, used slick shielding to protect fruit and nut trees from damage when working in them. The driver's protected position was below and behind the shielding, and the lights were mounted out of the way under the tractor's radiator bottom. This 1958 620 Orchard LPG was collected and restored by ardent John Deere collector Verlan Heberer of Belleville, Illinois.

1957 Deere 720 Diesel

When it was tested new in September 1956, the John Deere Model 720 Diesel Row-Crop was the largest and most powerful such tractor at the University of Nebraska trials. It produced 53.66 drawbar horsepower and set new fuel efficiency records. The 720 was also available as a high-crop or standard-tread tractor. Power steering helped keep this heavy tractor headed in the right direction. Chad Reeter of Trenton, Missouri, restored this pristine 1957 Model 720 Row-Crop.

1958 Deere 720 Diesel

This 1958 John Deere Model 720 diesel with wide front end had the power to pull a five-bottom plow under most conditions. With more than 50 horsepower available at its drawbar, the 720 was a productive machine. When fitted with wide multi-row cultivators, the big row-crop made quick work of crop cultivation. Some industry observers credit the 720 model as the reason John Deere took the lead in tractor sales in the late 1950s. Jim and Doris Finnigan of Shirley, Illinois, restored this nice 720 diesel wide front.

1958 Deere 820

The ultimate John Deere of its era was the "black dash" Model 820 diesel. With a black-painted instrument dash, the later 820 boasted 75.60 belt horsepower and 69.66 drawbar horsepower. The 820 could handle six 14-inch plows in average conditions and was popular in wheat country. The 820's big engine and six-speed transmission suited it to pulling wide hitches of tillage tools and grain drills. Former Deere mechanic and tractor restorer Don Ward of Chula, Missouri, owns this 1958 Model 820.

1960 Deere 330 Utility

John Deere launched its 30 Series tractors in 1959, which featured mostly cosmetic changes. The steering wheels and the tractor instrument panels were angled toward the operator in the new models. The sides of the tractors hoods also featured more yellow. Eventually, 30 Series tractors became popular with collectors because they were the last models of the John Deere two-cylinder era. This 1960 low-slung Model 330 Utility is a rare model of which only 247 were made. It belongs to Suzanne Burch of Sikeston, Missouri.

1959 Deere 530

Produced only as a row-crop tractor, the John Deere Model 530 came from the Waterloo tractor factory in 1959 with new flat-topped fenders with front-mounted dual lights. A deep-padded seat and an angled steering gear and instrument panel added to operator comforts. This 1959 wide-front Model 530 has front weights to help balance rear-mounted integral implements. It was a three-plow tractor in most soil conditions. Ken Smith of Marion, Ohio, owns this Model 530, among the last of the two-cylinder John Deere tractors.

1959 Deere 730 Row-Crop
Five-plow power and new features made the 730, John Deere's biggest row-crop machine, the most popular of the 30 Series. Some 17,000 of the 730 model were made and sold from 1959 to 1961. Electric starting, without the pony engine, was an option on the model. Gasoline, diesel, all-fuel, and LP fuel options were also available. The 730 was produced in row-crop, standard-tread, and high-crop variations. Steve and Lewis Schleter of Princeton, Indiana, use this 1959 730 diesel wide-front row-crop in their farming operation.

1959 Deere 730 Standard
Among the last of the John Deere two-cylinder tractors was this 1959 Model 730 standard diesel. Big clam-shell fenders and a fixed wide-front axle identify it as a standard-tread model. With 53.66 drawbar horsepower and five-plow capacity, it could "blacken" a lot of acres in a day. After U.S. production of the 730 ceased with the advent of the New Generation tractors in 1960, the 730 was exported as late as 1961. Professional restorer-collector Dan Peterman of Webster City, Iowa put this 730 diesel standard into tip-top shape.

1959 Deere 435 GM

Last announced of the two-cylinder John Deere tractors was the Model 435 diesel. It was powered by a General Motors two-cycle, two-cylinder 2-53 diesel. Its Nebraska tests showed 28.41 drawbar horsepower and 32.91 PTO horsepower at 1,800 rpm. The only John Deere with the GM diesel, it was built between March 31, 1959, and February 1960. This 1959 model is owned by professional tractor restorer and collector Ron Jungmeyer of Russellville, Missouri.

1960 Deere 830 Rice Special

Considered the ultimate in two-cylinder tractor development, the 1960 John Deere Model 830 was big and strong; it weighed 6 tons. This 830 Rice Special could pull 69.66 horsepower at the drawbar. Extra mud shielding and bearing protection around its brakes and rear axle area helped it withstand the rigors of the rice field. The deep-tread rice tires helped it dig its way through the mud. Front tires with their deep single rib kept it on course. Its worthy successor didn't come along until 1963, when the Model 5010 with 105.92 drawbar horsepower debuted. Ken Peterman of Webster City, Iowa, collected and restored this big gem.

Deere New Generation unveiling
On August 30, 1959, Deere & Company shocked even its own tractor dealers when it unveiled a whole new offering of four- and six-cylinder tractors at a conference in Dallas. Known as the New Generation, these models represented a significant change for the company that built its reputation on two-cylinder Johnny Poppers.

1960 Deere 3010 Diesel
John Deere's New Generation of Power, introduced in 1959, included the four-cylinder 3010 diesel. The 3010 diesel converted its 59.4 PTO horsepower to 54.5 on the drawbar. The new models completely replaced Deere's venerable line of two-cylinder machines. Not just updated from previous designs, the new tractors with all-new engines had 95 percent newly designed parts. This diesel 3010, owned by Ken Smith of Marion, Ohio, bears serial No. 1 in its model line. Smith also owns its 4010 diesel serial No. 1 counterpart.

1960 Deere 4010 Diesel
The big row-crop machine in John Deere's New Generation of Power was the 4010 diesel. Its six-cylinder diesel engine put out 84 PTO horsepower and 73.6 drawbar horsepower running at 2,200 rpm. The new model series brought power brakes to farm tractors and featured hydrostatic power steering, closed center hydraulics with up to three "live" circuits to drive power steering, and implement controls. Deere & Company invested $50 million in the development and tooling of the new tractor line. Styling of the new line again came from industrial designer Henry Dreyfuss of New York.

1962 Deere 4010 LPG High-Crop
Resembling a grasshopper ready to leap, this 1962 Model 4010 LPG High-Crop was the ninth of only seventeen made of this version. The high-crop option added "legs" to the 4010 tractor for use in tall-growing crops like sugar cane and staked tomatoes. The New Generation tractors included the 1010, 2010, 3010, 4010, and 5010 models. John Deere collector-restorer George Braaksma of Sibley, Iowa, transformed this tractor from its rusting remains that he found in Louisiana's sugar cane country.

1965 Deere 1010RS
The New Generation line also included smaller tractors like this 1965 Model 1010 RS. Designed for single-row cropping operations, the 1010 had its engine and drivetrain offset to the left for better one-row cultivation visibility. The 1010 four-cylinder gas engine provided 30.8 horsepower at its drawbar. It was a popular tractor in vegetable and tobacco production areas. Lewis Schleter of Princeton, Indiana, collected this 1965 Model 1010.

1964 Deere 1010 Crawler
The 1964 John Deere Model 1010 crawler was a heavy puller due to its superior traction and lower gearing. This small 30-horsepower crawler could pull three to four plows in most soil conditions. The 1010 replaced the Model 430. The 1010 crawler had a four-speed transmission and came with either five-roller or four-roller track suspension. Gas or diesel four-cylinder engines were available. Owner-restorer Randy Griffin of Letts, Iowa, specializes in collecting John Deere crawler tractors.

1965 Deere 2510
The Model 2510 row-crop tractor replaced the earlier Model 2010 in 1965. The 2510 was manufactured in Waterloo and was essentially a Model 3020 chassis with a Dubuque 2020 engine. Owner Laverne Schmidt of Klemme, Iowa, bought this diesel tractor new, used it for thirty-five years on his farm, then had it restored to like-new condition. It carries serial No.1 as the first diesel 2510 made. His 2510 row-crop is equipped with Roll-Gard and a canopy for roll-over protection and shade for the driver.

1965 Deere 4020

The Model 4020 is one of the all-time John Deere classics. It replaced the earlier 4010 model in 1964 and was very popular, accounting for 48 percent of all John Deere 1966 tractor sales in North America. Owner Tom Manning of Dallas Center, Iowa, hired out this 4020 for a role in the 1995 motion picture *The Bridges of Madison County*. Hogback Bridge near Winterset, Iowa, forms the background.

1964 Deere 3020

The new Power Shift transmission in the 1964 John Deere Model 3020 tractor allowed its operator to shift to one of eight forward speeds by using a single lever on the dash. Advanced hydraulic systems on the series provided power steering, power differential lock, and power for lifting and controlling attached implements. Along with the larger Model 4020, this 3020 is one of the all-time classic tractors, and collectors are busy restoring the forty-plus-year-old machines. Bruce Halverson of Huxley, Iowa, had this fine 3020 restored.

Chapter Six
Ford

1924 Fordson F
By 1923, the Fordson had sales of nearly 102,000 tractors, or 76 percent of all tractors sold that year. Ford was selling the tractor then for only $395 and winning the raging tractor price war. Henry Ford sold almost 740,000 Fordsons from 1917 to 1928, but not nearly the 13 million sales he made with his Model T Tin Lizzie. Ford collector Palmer Fossum of Northfield, Minnesota, restored this 1924 Fordson Model F.

N ot once, but twice Henry Ford changed the future of the tractor industry in the United States. Ford's first success in building tractors came with the legendary Fordson, launched in 1917. Then twenty years later, in 1939, he came out with the landmark Ford-Ferguson Model 9N. Both tractor models were milestone machines that drastically changed the way tractors were perceived, made, and used—not only in North America, but across the world.

Henry Ford, like most Americans of his era, was a farm boy. He was born in Wayne County, near Dearborn, Michigan, on July 30, 1863. As a boy and young man, he had hands-on experience working on his father William Ford's farm. He considered the work drudgery and dreamed of a way to make farm life easier and more productive.

Young Ford first began to experiment with steam engines while he was still on the farm. In 1893, while employed as an engineer with the Edison Illuminating Company in Detroit, Michigan, he started working on internal combustion engines, and in his spare time he made his first automobile. His Quadracycle, built with help from fellow Edison staffers, first rolled out of his small shed on June 4, 1896.

By 1903, Ford had organized the Ford Motor Company, and in 1908 the company started mass-producing the famous Model T automobile— a simple, inexpensive, and highly popular car that put America on wheels. By the end of its long and fruitful production run in 1927, a record 15 million Tin Lizzies had rolled off the Ford assembly lines.

Henry Ford wanted to work the same mass production magic on farm tractors. He built an experimental automobile plow in 1907, using one of his 1903 Model B car engines and transmission. More experimental tractors followed, and in 1915 Ford announced he would build a light two-plow tractor to sell for $200. Other entrepreneurs were listening.

Seeing gold in the Ford name, a Minneapolis, Minnesota, group organized the Ford Tractor Company in 1916, using the name of employee Paul B. Ford. The group then began producing a "Ford" tractor, hoping Henry Ford would pay to use the name on his own machine. Few of the awkward three-wheel, two-wheel-drive machines were made, but its presence caused Henry Ford to organize his tractor company as Henry Ford & Son in 1917 and to name his tractor the Fordson.

Orders came in for the Fordson even before it was built. In an attempt to keep food on British tables during World War I by promoting farm mechanization and more home food production, the British Ministry of Munitions ordered 6,000 Fordsons while its

1923 Fordson

At a time when other gas tractor makers were trying to compete with steam engines, Detroit's Henry Ford of Model T fame developed the simple two-plow Fordson in 1917. Its four-cylinder 251-cubic-inch engine had a 4x5-inch bore and stroke and produced about 20 horsepower at 1,000 rpm. It only weighed about 2,700 pounds and could pull two 14-inch plows at about 3 miles per hour. As first introduced, it was a no-frills machine. Robert Mashburn of Bolton, Mississippi, bought this 1923 Fordson tractor and completely restored it more than thirty years ago. Mashburn was driven in his restoration efforts by memories of his father working with the early tractor.

1910s Ford Automobile Tractor
Numerous firms across the United States offered conversion kits to turn the ubiquitous Ford Model T automobile into a farm tractor, including the Pullford Company of Quincy, Illinois. These kits ranged from rear-end conversion kits to full row-crop tractor makeovers.

construction and details of manufacture were still being worked out. Ford was able to deliver more than 250 tractors on that order in 1917, with the balance shipped to Britain in 1918.

Yet the U.S. farm implement business had a hard time taking Henry Ford and his little Fordson tractor seriously. The tractors of the day were primarily big ones, made to replace large lumbering steam engines. Even the "lightweight" tractors that were available usually weighed more than twice that of the Fordson.

THE DETROIT GROWLER

The 1917 Fordson was not only unconventional because of its weight. Its design also didn't fit with the then-standard three-wheel tractor configuration. Though it looks fairly conventional today with its standard-wheel configuration of large rear drive wheels and a smaller automotive-type steering gear in the front, the Fordson wasn't conventional at the time. It came equipped with a four-cylinder 251-cubic-inch inline gasoline engine, mounted lengthwise. Its 1,000-rpm 20-horsepower engine used Ford's Model T-type ignition with a flywheel-mounted dynamo supplying current to a vibrating high-tension coil on the engine

block. The engine castings were bolted to the transmission housing, producing a frameless unit, one of the first of its type.

The tractor's multiple-disk clutch ran in oil. Splash engine lubrication was augmented with oil circulated by use of built-in funnels and trays to distribute oil through the engine. Final drive from the transmission came through a worm gear in the enclosed differential housing to the enclosed axles, then out to the rear drive wheels. The Fordson weighed about 2,700 pounds, and its three-speed transmission gave it speeds of up to 6.8 miles per hour. It could plow at about 3 miles per hour. The little rig was considered a two-plow tractor.

Farmers had an easier time accepting the Fordson. They knew Henry Ford—he had put them on wheels via the cheap but tough little Model T. They knew Henry Ford's reputation for making good on his promises. They also soon knew where to buy a Fordson: the same place they bought their car. And even though Ford had to price his tractor at $750 instead of $200, the Fordson was still a hit.

When the Fordson was released to the domestic market in 1915, its sales took off immediately; 34,167 were sold in 1918,

57,000 more in 1919, and 67,000 in 1920. By 1922, Fordson sales were up to 69,000 units a year, accounting for 70 percent of all U.S. tractors sold. In 1923, nearly 102,000 Fordsons sold, making up 76 percent of all gas tractor sales in the country. That was the same year John Deere's Model D was announced and a year before IH's Farmall hit the market.

TRACTOR PRICE WAR

By 1921, the United States was in a post–World War I recession, and Fordson sales were down by 45 percent to only 36,793 tractors. Ford responded by cutting the price of the tractor three different times to a low of $395. His competitors also cut prices; the tractor price war was on. The farm implement leader of the day, International Harvester Company, not only cut prices for its 10/20 Titan tractor to $700, but threw in a free plow.

Ultimately, IHC bested the competition by developing a better product: a general-purpose tractor that soon eclipsed Ford and other makers in its all-around farm utility. The innovative machine was introduced in 1924 as the row-crop Farmall, another seminal tractor that, like the Fordson, moved farm mechanization a giant step ahead.

1919 Fordson advertisement

Ford Tractor Production
Henry Ford used the same mass-production assembly line techniques to produce his tractors as he did for putting together his automobiles. The automation allowed him to offer the Fordson at an astonishingly low price: $395. With such a deal, other tractor manufacturers were forced to lower the prices on their machines.

1928 Fordson F

The ubiquitous Fordson became even more useful with aftermarket products like this track conversion. Trackson Company of Milwaukee, Wisconsin, made the unit seen here on a 1928 Fordson F. Large-diameter wheel brakes were activated by the Fordson steering wheel to make it turn. The Fordson tractor encountered stiff new competition in the mid-1920s as International Harvester developed modern lightweight tractors. Fordson sales suffered, and in 1928 Ford moved its tractor production overseas to Cork, Ireland. Richard Stout of Washington, Iowa, bought this tracked Fordson in 1989 and invested a lot of time in returning it to operating condition.

The Fordson lost market share to the Farmall, as did most other tractors. By 1928, IHC was back in the lead with 47 percent of the U.S. tractor market, and in 1929 its sales made up 60 percent of U.S. tractor sales. Meanwhile, Ford sales plunged, and the company stopped making the Fordson in the United States in 1928, moving tractor production to Cork, Ireland. Ford was busy, though. It quickly converted Fordson tractor manufacturing facilities in Detroit over to production of the new Model A Ford automobile. Almost 740,000 U.S. Fordsons were made from 1917 to 1928.

1924 Fordson F
This Fordson has optional full fenders with built-in toolboxes. They helped keep the tractor from "turning turtle" and ending up on its back when its plow stuck on an immoveable object and reared up.

1929 Fordson N
Fordson tractors produced in Ireland were imported duty-free into the United States in relatively small numbers. This 1929 Irish Fordson Model N was an updated machine. Its engine displacement was increased to 267 cubic inches from 251 by widening the cylinder bore to 4.125 inches and raising its operating speed to 1,100 rpm. A high-tension impulse-coupled magneto replaced the old Detroit flywheel dynamo. The N produced 21 horsepower on kerosene and 26 horsepower burning gas. New cast front wheels replaced the old spoked wheels, and the tractor also had a stouter front axle. The road bands over its lugs make this tractor road legal. Some 31,471 Fordsons were built in Ireland from 1929 to 1932, after which Fordson production was moved to Dagenham, England. James Hostetler of West Liberty, Ohio, collected this tractor.

Many of the Irish-made Fordson Model Ns, with larger displacement engines equipped with high-tension magnetos, were imported duty-free into the United States. Large rear fenders and a new wheel design distinguished the Model N from the domestic Model F Fordsons. The Model Ns also had a water pump to help engine cooling.

The Fordson Model N was powered by the Fordson engine—boosted to 267 cubic inches of displacement by a bore increase to 4.125 inches. A conventional magneto replaced the Model T flywheel dynamo, the operating rpm was raised to 1,100, and the engine was available as a kerosene or gasoline version. The Model N also gained weight in the transition and now weighed in at 3,600 pounds.

In 1933, Fordson production shifted from Cork to Dagenham, England, where it continued until 1938. In 1937, the first Fordson row-crop tractor, the tricycle All-Around, was introduced by the English Ford Motor Company, Ltd. The All-Around tricycle-gear row-crop was exported to the United States in an attempt to earn back some sales lost to the other general-purpose tractors. By the end of Fordson production in England in 1938, another wave of Ford's ingenuity was about to wash over the world. This time Henry had help from abroad, an Irish inventor named Harry Ferguson.

A GENTLEMAN'S AGREEMENT

Both Henry Ford and inventor Harry Ferguson had long dreamed of making implements integral with the tractor, avoiding the tacked-on way of hitching implements. That practice had carried forward to tractors after they started replacing horses. Many farmers simply modified their horse-drawn equipment for tractor use by sawing off most of the long wooden tongue and attaching a new strap iron hitch.

Harry Ferguson had worked on his integral hitch for about twenty years before he demonstrated his Ferguson System to

1937 Fordson N All-Around
The Fordson row-crop tractor was the English-made 1937 tricycle-configured Model N All-Around. The All-Around was painted this deep blue with orange trim. The Model N had been upgraded in 1935 with the steering column angled at about 40 degrees and an optional PTO added. Rubber tires and a starter and lights also became options. Earl Resner of Grand Junction, Colorado, restored this 1937 All-Around. Front-mounted cultivators were used on the tractor.

1937 Fordson N All-Around
Wide-tread rear axles with big wheels and a tricycle-configured front mark the basic changes that made the All-Around a row-crop tractor. Fordson also included outboard brakes for the rear wheels on the model.

Henry Ford at Ford's Fairlawn, Michigan, farm in the fall of 1938. Ferguson's tractor with its integral three-point hydraulically controlled hitch worked well, and Ford was impressed. He wanted to buy Ferguson's patents on the spot. But Ferguson indicated they weren't for sale at any price. So they made an agreement—an oral agreement sealed with just a handshake.

According to Ferguson's later recollection, the agreement was that Ferguson would be in charge of design and engineering of the hitch, Ford would build the tractor and assume those risks, Ferguson would distribute the tractors wherever and however he pleased, and either party could end the agreement at any time for any reason.

THE 9N DEBUTS

In June 1939, the new Ford tractor with the Ferguson System was ready. The Model 9N boasted a 120-cubic-inch, four-cylinder high-compression gasoline engine rated at 28 horsepower at 2,000

1938 Fordson N
The Fordson became a modern-looking standard-tread tractor by 1938, when this Model N was made in Dagenham, England. Orange was the Fordson color briefly until 1939, when Fordsons became green to make them less visible from the air during World War II when England was under aerial attack.

1938 Fordson N

Rubber tires aided this Fordson's productivity. Notches along the radiator sides held the radiator curtain for engine heat adjustment. Richard C. Vogt of Enid, Oklahoma, took this tractor apart down to the last nut when he completely restored it. His father had owned a 1937 Fordson, and Richard wanted one like it.

Ferguson Plow and Three-point Hitch
The Ferguson plow and three-point hitch on the Ferguson-Brown tractor is similar to the one Harry Ferguson used to successfully demonstrate his Ferguson system to Henry Ford in 1938.

rpm. Of standard-wheel configuration, the tractor had an adjustable-width front axle, and its rear wheels could be reversed on dished rims so the tractor could straddle two crop rows at an 80-inch setting. It also was a row-crop design.

Rubber tires were standard, as was the generator and starter. Also standard was the PTO and the Ferguson System of hydraulics for control and lift of three-point mounted implements. The 9N even came with a muffler in the under-mounted exhaust system. Only the electric lights, rear-mounted two-bottom plow, and row-crop cultivator were optional.

In the 1940 Nebraska tests, the Model 9N produced 16.31 horsepower on the drawbar and 23.56 on the belt. Its 3.1875x3.75-inch four-cylinder gas engine hummed at 2,000 rpm. It managed a pull of 2,146 pounds.

1936 Ferguson-Brown Type A
An English-made 1936 Ferguson-Brown Type A tractor, similar to this one, was used in October 1938 by Harry Ferguson to demonstrate his Ferguson system of plow-mounting and hydraulic control to Henry Ford. As a result of that demonstration, Ford agreed to make a tractor incorporating the Ferguson system that Ferguson would sell. The tractor that emerged from their informal agreement was the Ford-Ferguson Model 9N of 1939. It used no parts of the Ferguson-Brown tractor. David Lory of Platteville, Wisconsin, collected this English-made tractor to add to his Ford tractor collection. Only 1,350 of this model were made in England from 1936 to 1939.

The Ford-Ferguson, as it came to be known, was a thoroughly modern tractor, from its streamlined radiator grille to its three-point hitch. It was not in any sense just a redo of the Fordson of old. About the only thing they had in common was their gray paint.

Comfortably straddling the transmission housing like mounting a horse, the operator sat forward of the rear wheels for a smoother ride and a more intimate view of his work. To the right was a small lever controlling the hydraulics. Stops were provided on the lever guide for implement depth settings. Operating the tractor was so easy that a child could do it—and many did.

Two trailing lift arms hinged from the bottom of the axle were linked to the built-in hydraulic lift arms on the upper part of the transmission. The third point of implement attachment was to a high-mounted centered link that worked in concert with the lift arms. Ferguson's integral hit worked as if the three points of implement attachment converged near the center of the tractor's front axle. That let the implement trail, as if it was attached at the front axle center.

The 9N's new concepts led other manufacturers to imitate and improve on those features for the next twenty years. "The

1939 Ford-Ferguson 9N
Considered the father of the modern farm tractor, the 1939 Ford-Ferguson Model 9N was a compact utility row-crop machine with integral hydraulics operating a unique three-point implement attachment. Leading the industry, it provided draft sensing to supply both up and down pressure to the mounted implement. Rubber tires, electric starting, and PTO were standard. More than 78,000 were sold from 1939 to 1941. Ford tractor specialist Roger Forst of Prairie du Chien, Wisconsin, restored this beauty now rapidly approaching its seventieth birthday.

combined results of the Ferguson implement linkage and hydraulic control mechanism were a revolutionary new ease of implement operation," wrote R. B. Gray in his 1975 book, *The Agricultural Tractor: 1855–1950*. "The whole trend of design of farm tractors and equipment in ensuing years has been strongly influenced by the development of the Ferguson System."

Sales of the handy little tractor were brisk. By the end of 1940, more than 35,000 Ford 9Ns had been built, and in 1941 nearly 43,000 more came off factory lines. That all changed abruptly in 1942, as war efforts shifted to making defense products and tractor production was severely limited. After the war, the 9N's production resumed at a brisk pace. Between June 1939 and July 1947, more than 300,000 Model 9N and 2N Ford-Fergusons were built.

The Model 2N was introduced in late 1942. It included improvements that helped the operator and added to safety. The left brake pedal was moved to the right side with the right brake pedal so the operator could disengage the clutch with his left foot while braking both rear wheels with his right foot. On the 9N, the left wheel brake and clutch pedals were both on the left, making it impossible to operate the left brake and the clutch simultaneously, as might be needed in stopping during a turn at the end of the field. The 2N also included a lockout on the starter that prevented its operation if the transmission was in gear. Other changes included the elimination of chrome trim on the tractor. With rubber rationing during World War II, some 2Ns left the factory with steel wheels.

THE HANDSHAKE AGREEMENT ENDS

Postwar recovery put stresses on tractor manufacturers to make up for lost time and meet the heavy demand for machines. War price controls had resulted in massive losses on the production of the

1939 Ford-Ferguson 9N
The business end of Ford's 1939 Model 9N Ford-Ferguson includes the hydraulically operated lift arms exiting the transmission case just rear of the tractor seat. The tractor's horizontal drawbar element here is installed in the two main implement attachment points of the three-point system. The third point of attachment is at the locking pin centered above the differential case.

Ford-Ferguson System of Implements advertisement

1939 Ford-Ferguson 9N
Freeman Miles, retired Ford tractor dealer from Dothan, Alabama, took mules in trade for tractors like this 1939 Ford-Ferguson 9N. Then he taught the new owner how to drive the tractor. Miles remembers selling the tractors for $575. This tractor was the first of its kind sold in Alabama. Miles completely restored it. His tractor has the mounted 14-inch two-bottom plow. The cast rear-axle hubs with their dished centers are original. Ford stylists made the tractor an appealing practical machine with its Art Deco touches.

1939 Ford-Ferguson 9N
Freeman Miles' 9N model has a generator and starter, but no lights. Its vertical stamped steel grille replaced the earlier horizontal cast aluminum grille.

1939 Ford-Ferguson 9N
The 9N's implements were lifted for transport with the same hydraulic system that controlled them in operating position. The lever for the hydraulic control was located within easy reach of the operator's right hand, just to the right of the tractor seat.

1952 Ford 8N
Ford and Ferguson went their separate ways in 1946 after their 1939 verbal agreement failed. Ford designed its own Model 8N tractor and produced it from 1947 until 1952. This is a 1952 red belly 8N, equipped with the Ford 119.7-cubic-inch four-cylinder engine, rated at 21.95 horsepower at 2,000 rpm. It has a four-speed transmission and weighs about 2,000 pounds. Roger Forst is its second owner. His father Leonard bought the 8N new, after one of the horses in his last team died. Its serial number is 517842. Ford made more than 524,000 Model 8Ns.

1953 Ford NAA

The Ford NAA Golden Jubilee model replaced the Model 8N in 1953, as the Ford Motor Company celebrated its fiftieth anniversary. The Ford Red Tiger 134-cubic-inch four-cylinder overhead valve engine rated at 31.14 horsepower at 2,000 rpm. The NAA had live PTO, new hydraulics, and other changes to end patent claims from Harry Ferguson, who had engineered the system used on the original 9N Ford-Ferguson. The NAA model was produced in 1953 and 1954. Only the 1953 model included the Golden Jubilee medallion on the radiator grille. William Krause of West Des Moines, Iowa, owns this landmark Ford.

Ford tractor. No price increases were allowed during the war, so Ford must have felt those pressures in 1946 when it tried to renegotiate its agreement with Ferguson. Ford wanted to either buy him out, or buy into his distribution rights. But the parties couldn't come to terms, and the 1938 gentleman's agreement effectively ended on December 31, 1946. Ford-Ferguson was no more, but Ford agreed to make tractors for Ferguson to sell through June 1947.

Ford quickly established its own distribution firm, Dearborn Motor Corporation. As it ended production for Ferguson in June 1947, Ford announced its own new tractor, the Model 8N. Little had changed on the 8N except for its colors and some Ford

1953 Ford NAA
Engine instruments, including a tachometer/hour meter, helped the operator keep tabs on the NAA. The NAA Ford was a bigger, stronger tractor than the previous Model 8N.

Right: **1958 Ford 901 Powermaster**
The tricycle row-crop Model 901 Powermaster joined the Ford tractor lineup in 1957. It featured the 172-cubic-inch four-cylinder engine and was a full three-plow tractor. It produced 50.2 PTO hoursepower and 43.28 drawbar horsepower. In 1955, Ford became more than a one-tractor company. The Models 600 and 800 gave Ford two-plow and three-plow power sizes in the utility tractor configuration. In 1956, Ford offered the 700 and 900 Series in tricycle-type row-crop models. Segmented wheel weights helped to ballast the tractor for heavy pulling. Workmaster Fords used the 134-cubic-inch engine and the Powermasters had the 172-cubic-inch engine. William R. Van Zante of Pella, Iowa, bought and restored this Ford.

1958 Ford 901 Powermaster
Left and on next page: Red grille
elements and a red strip down the
top of the hood marked the restyling
of the 1957 Ford tractors, while the
basic three-point hitch and hydraulics
remained similar to those
of twenty years earlier.

improvements. The tractor engine, drivetrain, and axles were painted red, and the hood, fenders, and wheel rims were a warm light gray. Ford's red-painted script trademark was stamped into the fenders and front hood. The 8N had a four-speed constant-mesh transmission, better brakes and a better steering gear, and an improved starter lockout. And it had basically the same hydraulic three-point hitch and implement control system as before. Patent infringement? Royalties? Ford made no payments to Ferguson for these, so he sued Ford for $251 million.

Meanwhile, Harry Ferguson was busy on the manufacturing front, too, going it alone in the tractor business. His Model 2N look-alike was the new Ferguson TE-20, assembled in Coventry, England. Soon after the introduction of the TE-20, the Model TO-20 Ferguson tractor was built at a daily rate of 100 units in Detroit, Michigan.

Ferguson's lawsuit against Ford dragged out for four years, and when it was finally resolved, a worn-down Ferguson settled for only

$9.25 million. Some of the Ferguson patents had expired by the time the suit was resolved. Ford had to redesign only a few of his parts to continue producing tractors with the now-famous built-in hydraulically controlled three-point implement hitch system.

FORDS OF THE FUTURE

To mark the Ford Motor Company's fiftieth, or golden, anniversary in 1953, Ford brought out the new NAA "Jubilee" model. It came with Ford's first valve-in-head tractor engine. The new Red Tiger engine had a displacement of 134 cubic inches and turned out 31.14 horsepower at 2,000 rpm, giving the Jubilee enough power to pull three 14-inch plows. A new grille design and treatment distinguished it from the 8N. Just slightly taller and longer than the 8N, the Jubilee was produced only between 1953 and 1955.

Ford got out of its single tractor model rut in 1955 when it brought out five new tractors in two power sizes. The 600 and

800 model series were three- and four-plow tractors available in gasoline and LP versions. The series included the Model 640, 650, and 660 three-plow power tractors, and the 850 and 860 four-plow machines.

In 1958, the first Ford tractors with American-made diesel engines were the Models 801 and 901. By 1959, Ford's Powermaster and Workmaster Series arrived with a choice of gasoline-, diesel-, or LP-fueled engines. The push for power led Ford to introduce its largest tractor to date in 1961: the five-plow six-cylinder diesel Model 6000. It produced 66 horsepower.

By Henry Ford's death at the age of eighty-three on April 7, 1947, some 1.7 million tractors bearing his name had come from Ford factories. Ford's mass-production techniques, including his fabled assembly line, taught the world how to make tractors as well as cars. Vast changes in the intervening years since Henry Ford's lifetime have further changed tractors, but also changed where and by whom they are made and how they are marketed.

In 1985, Ford combined its tractor operations with those of New Holland by buying the Pennsylvania company from the Sperry Corporation. Ford-New Holland then was sold by Ford to the Fiat Group of Italy in 1990. Now, the familiar Ford oval trademark with the script name has disappeared from tractors. Block letters spelling Ford were used only until 2001, under agreements reached with Fiat when it purchased Ford-New Holland.

Today, New Holland makes and sells tractors through CNH Global, a unit of the Fiat group of Italy. CNH, the second-ranking farm equipment company, also owns the Case-IH farm and industrial equipment line.

Chapter Seven
International Harvester

1936 Farmall F-30
The Farmall's big brother, the Model F-30, was introduced in 1931. Scaled-up from the original Farmall, the Farmall F-30's four-cylinder engine produced more power for the tractor—enough to handle a four-row cultivator or a three-bottom plow. The engine could be set up to burn gasoline, kerosene, or distillate. Harold Glaus of Nashville, Tennessee, grew up with this 1936 Farmall tractor that his parents bought just before he was born. The rear tires are original.

Cyrus Hall McCormick lived the American dream. He rose from being an ordinary Virginia farmer to a wealthy industrialist and founder of The International Harvester Company after taking over a design for a grain reaper from his father, Robert McCormick Jr.

Cyrus was just twenty-two when he went to work developing the machine his father had started in Rockbridge County, Virginia. By 1831, he had a working reaper to demonstrate and three years later, Cyrus patented the machine. In 1840, he made and sold his first reaper: a humble start to what eventually became the vast International Harvester Company of Chicago, Illinois.

The McCormick reaper was a relatively simple machine pulled by one horse, but it was a major step in revolutionizing the grain harvest. It eliminated the back-wrenching human toil of swinging a scythe or cradle to cut grain. McCormick's reaper could cut three times as much wheat in a day as a good man with a scythe. Later on, binders could cut and bind seventeen times as much wheat in a day as a farm worker. When coupled with the larger new threshing machines that soon developed, human productivity was multiplied many times, and food production soared.

Like other farm machine developments of the day, McCormick's reaper had lots of company—other inventors and makers pursuing the goal of the perfect harvester. Some of their efforts did shape the way grain was harvested, as well as form the companies that later made harvest equipment.

Once proven successful in Virginia, McCormick headed west with his reaper in 1847 for the promise of larger markets in the awakening prairies. Like other pioneering farm equipment makers who moved to the Midwest during that era—including John Deere and J. I. Case—McCormick was right in his pick of geography and timing. He built his reaper plant in Chicago, a location that was later to become home to the huge International Harvester Company.

Reapers flowed out of the plant between 1848 and 1850, 4,000 to be exact. Up to that time, McCormick had made only about 1,200 reapers. His brothers Leander and William S. McCormick soon joined him in the reaper manufacturing business and played big roles in its success. With the continuing popularity of the reaper, all three prospered. William died first in 1865, and in 1879 the partnership was incorporated as McCormick Harvesting Machine Company. Cyrus died in 1884, and in 1890 his widow, Nancy, and son, Cyrus Jr., bought out brother Leander's shares. Cyrus Jr. successfully led the company for many years after his father's death.

The 1871 Chicago fire destroyed the McCormick reaper plant and left the firm with a $600,000 loss. Vital records, fortunately, were recovered intact. The factory was replaced in 1872 and 1873 on a new, larger site as a bigger, more modern plant.

1911 International Harvester Titan Type D
International Harvester tractors were well established by 1911 when this Titan Type D model huffed out of the Milwaukee Works shops. Using the one-cylinder IHC Famous engine, the 25-horsepower rig had a bore of 10 inches with a 15-inch stroke and ran at 290 rpm. Ten-inch steel channel iron made up the frame for the 9-ton behemoth. Hopper evaporative cooling helped the engine stay cool. IHC collector Leroy Baumgardner Jr. of Hanover, Pennsylvania, shows his engines at the Rough & Tumble Show at Kinzers, Pennsylvania.

1911 International Harvester Titan Type D
IHC made 1,000 of these 25-horsepower Type D Titans in 1911.

By buying patents that showed promise and innovating those concepts, as well as negotiating manufacturing rights from others, the McCormick Harvesting Machine Company was able to keep up with fierce competition from rival firms. But there was one aggressive manufacturer bearing down on them right there in Chicago.

In 1870 William Deering, a wealthy dry goods wholesaler from Maine, invested in an Illinois reaper called the Marsh harvester and helped it to succeed. It was first built by brothers Charles W. and William W. Marsh of Shabbona Grove, Illinois,

and was patented in 1858. The Marsh machine was a forerunner of the binder—an advanced reaper that not only cut and bunched the ripe grain, but bound it neatly into a bundle.

By 1880, William Deering was the sole owner of the Deering Harvester Company, and he was selling twine-tie binders that combined the strengths of the Marsh harvester with those of the Appleby twine binder. Within the next decade, the McCormick and the Deering firms were both on top as harvester providers and were beginning to compete for sales of related harvesting equipment. Deering tried to sell out to McCormick in 1897, but financing couldn't be arranged. Both firms had strained themselves financially in keeping up with the other.

A MATCH MADE ON WALL STREET

A merger plan to put McCormick and Deering together, and solve their competitive dilemma, was worked out in 1902 in the New York offices of J. P. Morgan. The new International Harvester Company was formed by combining the assets of McCormick Harvesting Machine Company of Chicago; Deering Harvester Company of Chicago; Plano Manufacturing Company of Plano, Illinois; Milwaukee Harvester Company of Milwaukee, Wisconsin; and Champion Reaper Works. The new company was considered huge for the time, with a $120 million capitalization.

The new IHC combined the best features of the harvesting equipment it had acquired and moved forward into the twentieth

1912 International Harvester Mogul Type C
More than 2,440 of the IHC Mogul Type C gas tractors were sold between 1909 and 1914, its last production year. The single-cylinder engine of 10x15-inch bore and stroke was rated at 25 horsepower at 240 rpm. Hit and miss governing fired the engine only enough to keep up its speed. Forward drive was through massive spur gears, but reverse used friction drive. Cooling water flowed over the screen wire and back into the tank. IHC sold the Mogul and the Titan gas tractors through separate dealerships. Green-painted Moguls were sold by McCormick dealers and gray-painted Titans by Deering dealers. Paul Stoltzfoos of Leola, Pennsylvania, restored this rig shown at Kinzers, Pennsylvania.

century to capture other markets. Its foreign trade grew rapidly in Europe, the British Empire, South America, Africa, and Russia. A new factory was built in Hamilton, Ontario. New equipment lines were added when IHC bought their manufacturers. In 1903, an old rival, the D. M. Osborne Company, was bought. Then the Weber Wagon Company of Chicago was added, as well as Aultman-Miller of Akron, Ohio. Tillage, haying, and other lines came with the purchase of the Keystone Company of Rock Falls, Illinois, in 1904. IHC's first European plant was built at Norrkoping, Sweden, in 1905. Other plants in Germany, France, and Russia followed. International truly *was* international. Amazingly, more was to come.

TIME FOR TRACTORS

A powerful rhythmic pounding sound was starting to echo over the prairies, the noise of the newfangled gasoline traction engine. IHC heard it as another opportunity and responded with a machine of its own, well, partly. In 1906, International put one of its single-cylinder, 15-horsepower, Famous horizontal gasoline engines made at the Milwaukee Works onto a Morton-made friction-drive truck, or chassis. The pairing worked, so International made more of these early tractors for testing that year and another 200 in 1907. Then between 1907 and 1911, more than 600 Type A tractors with 12-, 15-, or 20-horsepower engines were made. During that time, the Type A's friction drive gave way to gears, except for friction drive reverse.

1912 International Harvester Mogul Type C
The tractor operator had to really turn the steering wheel to maneuver the Mogul's nearly 9 tons.

1917 International Harvester Titan 30/60
Two cylinders, with a bore and stroke of 10x14 inches, helped the 1917 IHC Titan 30/60 achieve its admirable horsepower rating. The Titan 30/60 could pull a ten-bottom plow at 2 miles per hour, or it could belt up to a big threshing machine and drive it with its big belt pulley. The 30/60 Titan was started with a small air-cooled engine driving the big engine's left flywheel through friction contact. Such large tractors were soon replaced by lighter machines. IHC fans Mary and Wendell Kelch of Bethel, Ohio, bought this 30/60 in Montana in 1998 and had it in showroom condition a year later. They figured it took 3,000 hours of work and 30 gallons of paint to bring it back.

1917 International Harvester Mogul 10/20
Hopper cooling under the round stack was all that this 1917 Mogul 10/20 needed for cooling. The single-cylinder 8.5x12-inch engine gave the 10/20 enough power, through a two-speed transmission, to pull a three-bottom plow. The tractor's open chain drive powered only the furrow (right) wheel. In 1989, brothers Howard and Roger Schnell of Franklin Grove, Illinois, restored the old tractor their father owned nearly eighty years ago.

A Type B machine with a 20-horsepower engine and a full-length rear axle was offered next, but only 301 were made. By 1909, a two-speed Type A with 12 horsepower was available, but only sixty-five of those were put together. Though sales figures for the Type A and B were modest by any measure, IHC was already leading the industry in sales of gasoline traction engines by 1911.

IHC had kept separate dealerships after its 1902 merger and tried to give Deering dealers one tractor line to sell and McCormick dealers another. The green-painted Moguls became the McCormick offering, while the gray-painted Titans were at Deering dealers. The year the 45-horsepower Mogul came out, 1910, IHC added tractor production to its Chicago plant. The Mogul was powered with a two-cylinder horizontally opposed engine rated at 345 rpm. In 1911, a large but similar Mogul 30/60 was introduced. In 1911, the two-cylinder Titan was first made as a 45-horsepower tractor, then upgraded to a 30/60 rating with a larger engine.

In 1912, a 15/30 Mogul went into production. It was a one-cylinder version of the 30/60 Mogul. The 15/30 engine turned at 400 rpm. The Mogul introduced in 1913 was rated at 12/25

***Above and opposite*: 1919 International Harvester Titan 10/20**
Chief competitor for sales of the green Mogul 10/20 was the gray Titan 10/20. Both were built by IHC. This 1919 Titan was the more popular of the competing IHC machines because its two-cylinder side-by-side engine provided smooth, steady power. The front-mounted tank held circulating cooling water for the engine. Belt pulley work was done on the right side of the Titan. Its early success made the name Titan nearly synonymous with tractor. IHC built more than 78,000 Titans from 1916 to 1922, during the price war with the Fordson tractor. Earl Scott of Marysville, Ohio, put this 10/20 in his eclectic tractor collection.

horsepower from a two-cylinder horizontally opposed engine. The models were heavy monsters and looked more like steam engines than gas tractors.

IHC'S FIRST FOUR AND NEW LIGHTWEIGHTS

A Titan four-cylinder (or double twin) horizontal crossmount engine tractor joined the line in 1914. Begun as a 12/25, the Titan was upgraded to a 15/30 by 1918. It was a four-plow tractor weighing about 8,700 pounds. Its two-speed spur-gear transmission gave it operating speeds of 1.9 miles per hour and 2.4 miles per hour at an engine speed of 575 rpm. Heavy roller chains transmitted power to the 5-foot-diameter rear wheels. Many Titans were equipped with an enclosed cab, and the tractor sported a cell-type front-mounted radiator cooled with a belt-driven fan. The tractor was manufactured in Milwaukee between 1915 and 1922 and its various versions sold about 5,500 copies.

The most popular IHC tractor to that date arrived in 1914: the lighter-weight two-plow Mogul 8/16. The Mogul 8/16 had a one-cylinder, 8x12-inch, 400-rpm horizontal engine with hopper cooling. It had one speed forward, for 2 miles per hour, and a reverse.

Next page spread: **1917 International Harvester 8/16 and 1916 International F Truck**
The International 8/16 was the first tractor offered in North America with optional power take-off or PTO. A powered shaft at the rear of the tractor provided engine power to drive a trailed implement like a binder. To compete with the Fordson, IHC sold the 8/16 in 1922 with a free plow for $670. In its working life, the 1916 International Model F truck pictured here delivered fuel to farms in northwest Ohio. A Model F truck similar to this one was the first truck to climb Pike's Peak in Colorado. Merrill Sheets of Delaware, Ohio, collected these vintage IH machines.

Narrow spacing of the two front wheels, which were suspended from a gooseneck frame, allowed for shorter turning than on previous tractors. The Mogul 8/16 was built from 1914 through 1917 with more than 14,000 units made—more than any of the previous ICH models combined. It made a strong argument for smaller tractors.

A slightly larger and more powerful 10/20 Mogul replaced the 8/16 in 1916. Its bigger 8.5x12-inch single-cylinder engine was coupled to a two-speed gearbox. The 10/20 also had fenders. Only about 9,000 of the 10/20 Moguls were made, perhaps due to the popularity of IHC's other 10/20—the two-cylinder Titan available through Deering dealers.

The Titan 10/20 was introduced in 1915. It had a twin-cylinder, side-by-side, 6.5x8-inch horizontal engine running at 500 rpm. Thermosiphon circulation transferred engine heat to a front-mounted cooling tank where heat was lost by radiation. Other features included automobile-type steering of the two narrow-mounted front wheels and two forward speeds of 2.25 and 2.875 miles per hour, with a reverse of 2.875 miles per hour. It was a relatively lightweight tractor for its day at about 5,700 pounds. The Titans were painted gray with red wheels.

More than 78,000 of the 10/20s were made between 1915 and 1922, making it the new IHC tractor sales leader. Unlike the

1917 International Harvester 8/16
A modern four-cylinder overhead-valve truck engine and a nearly modern tractor chassis combined to make the 1917 International 8/16 tractor. The 4x5-inch engine turned over at 1,000 rpm to make its rated 8/16 horsepower. Some 33,138 of the 8/16 were made, but it couldn't compete with the Fordsons pouring out of Detroit. However, the groundwork the 8/16 laid helped IHC build tractors in the 1920s that put IHC back on top. Marylyn and Merrill "Pinky" Sheets of Delaware, Ohio, collected and restored this historic tractor.

"pop-pop-pause, pop-pop-pause" rhythm of the John Deere two-cylinder machines, the Titan 10/20 had a regular exhaust note due to its single-throw crank and regular power strokes every 360 degrees of crank rotation. With pistons and cranks both changing directions simultaneously on every stroke, the Titan 10/20 created its own characteristic set of vibrations at some throttle settings.

The 10/20 might have sold even better but for the competition brought on by the Fordson, marketed from 1918 on. In the tractor price wars of the 1920s, Ford cut the Fordson's price in 1922 to $395. IHC responded by reducing the price of the Titan 10/20 to $700, and it added a three-bottom P & O plow to sweeten the deal; IHC had recently purchased the Parlin & Orendorf Plow Company of Canton, Illinois.

GOING MODERN

The new IH Model 8/16 of 1918 was almost streamlined in appearance. The engine housing tapered toward the front—where there was no radiator grille. The 8/16 was powered by a vertical

1928 McCormick-Deering 10/20
Two new IHC tractor models helped turn the tide in the tractor price wars of the 1920s. The modern gear-drive Model 15/30 of 1921 and the smaller Model 10/20 of 1923 (above) were built to overtake the Fordson. The 10/20 and 15/30 were of similar construction with a one-piece cast frame running the length of the tractor. All drive components were enclosed. Replaceable cylinders and ball-bearing crankshafts were features of what became McCormick-Deering tractors. Clem Seivert of Granger, Iowa, rebuilt this 1928 Model 10/20.

four-cylinder 4x5-inch bore-and-stroke engine mounted lengthwise inline on the channel frame. Similar to one in the IHC Model G truck, the engine revolved at 1,000 rpm. Engine cooling, as on the truck, was provided by a fan-blown radiator placed behind the engine. The engine had removable cylinder sleeves and high-tension magneto ignition with impulse starter. The engine was coupled to the three-speed transmission with a multiple-disc dry clutch. Final drive was with roller chains to sprockets inside the rear wheels. The weight of the tractor was trimmed to about 3,300 pounds. It was rated as a two-plow tractor and the first to offer power take-off (PTO) to power implements.

The 8/16 was produced as the VB, HC, and IC Series from 1917 through 1922. The IC Series had a larger 4.5x5.5-inch bore-and-stroke engine. According to available serial numbers, more

1928 International Harvester 10/20
The McCormick-Deering Triple-Power Tractors produced drawbar, pulley, and PTO power. PTO was optional, but it was designed into the tractor and could be added later. Many implements made by IHC were PTO operated. By the end of 10/20 production in 1939, more than 215,000 of the two-plow tractors had been built.

than 33,000 of the model were produced with the later, more powerful version selling more than 16,500 units. It too had its price cut to $670, with a free two-bottom plow, to compete with the Fordson during the fierce price competition.

The "modern" tractor was further defined with the 1921 introduction of the new McCormick-Deering 15/30 Gear Drive Tractor. Replacing the aged Titan 15/30 of 1914, the new Model 15/30 was of one-piece cast-iron-frame construction; it added power take-off capabilities and featured a new International four-

The Powerful New McCormick-Deering 15-30

A Few 15-30 Features

McCormick-Deering high-tension magneto ignition.
New manifold design, increasing fuel efficiency.
Protected air supply.
Circulating splash engine lubrication.
Filtered fuel supply.
Efficient kerosene carburetion.
Friction-free ball-bearing crankshaft.
Three forward speeds.
Accessible construction.
Removable cylinders.
Replaceable parts throughout.

THE POWER in the new 15-30 McCormick-Deering is the symbol of profit farming on a comfortable, efficient scale. With this powerful perfected tractor special opportunities lie ahead of you. Its owner is equipped to rise above the old cramped style of farming—to take full advantage of man-power, acreage, crop, and season—to cut to the bone the production costs that eat profit away—*and to build for future expansion.*

This is a *McCormick-Deering* tractor. So you may be positive that its liberal power is matched by new improvements and refinements **all** along the line. The 4-cylinder power plant, clutch, transmission and differential assemblies, built into a rigid 1-piece main frame, give great reserve strength. All important wearing parts run in a bath of oil. Ball and roller bearings at 34 points add to easy running and long life.

Considering ample power, flexibility, long life, economy, price, service, and easy operation with the equipment which is as important as the tractor itself—here is the tractor of tractors. Ask the dealer about the new 15-30 McCormick-Deering. Other McCormick-Deering Tractors—the 10-20 and the all-purpose Farmall. Catalogs on request.

INTERNATIONAL HARVESTER COMPANY
606 So. Michigan Ave. of America Chicago, Ill.
(Incorporated)

A Little Story of Power Farming

This picture is drawn from a photograph of W. A. Asmussen, of Agar, S. D., riding his new 15-30 McCormick-Deering tractor. Mr. Asmussen is farming 1700 acres of land with the same crew (himself and two men) with which he farmed 600 acres three years ago using 18 horses. Now he has the 15-30 and a Farmall and modern power farming equipment. Note the special lamp mounted on the fender, permitting night work in rush seasons. He says he hasn't had a horse hitched up this year. They handle the 1700 acres easier and have more leisure than on the smaller acreage.

McCORMICK-DEERING

1920s McCormick-Deering 15/30 advertisement
McCormick-Deering's Model 15/30, introduced in 1921, had a one-piece cast-iron frame and a fully enclosed drivetrain with gear-driven rear wheels. It featured a new International four-cylinder engine with ball-bearing main bearings and power take-off capabilities. By the end of the three-plow 15/30's production, 157,000 had rolled off company lines.

cylinder engine with ball-bearing main bearings and removable cylinder sleeves. The new 15/30 also had a fully enclosed drivetrain with gear-driven rear wheels. Wheel placement was configured in what is now called standard configuration. Its vertical four-cylinder engine was positioned lengthwise, and it had a bore and stroke of 4.5x6 inches, and operated at 1,000 rpm.

When the 15/30 Gear Drive was upgraded in 1929, the earlier version had already sold nearly 100,000 units. The new Model 15/30 (known later as the 22/36) was a much-improved machine. IHC increased the bore of the engine to 4.75 inches, and its operating speed was boosted to 1,050 rpm. The 16/36 sold another 57,500 tractors from 1929 until its phase-out in 1934.

IHC introduced its replacement for the International Model 8/16 in 1923. The new 10/20 was a smaller version of the 15/30 with similar features, including the one-piece cast-iron frame and built-in PTO for powering drawn implements. The two-plow 10/20 became one of the most popular tractors of its time with more than 215,000 sold between 1923 and 1939. Its production peaked in 1929 when nearly 40,000 sold. The 10/20 was powered with another smooth-running four-cylinder IHC engine. Nebraska tests showed the 10/20 produced 10.9 drawbar horsepower and 20.46 belt horsepower in 1927. Narrow-tread and orchard versions were made beginning in 1925. Rubber tires became a welcome option in the mid-1930s.

THE FARMALL REVOLUTION

Most of the tractor companies, including IHC, developed and built very lightweight motor cultivators starting in about 1915. International's entry to the field in 1916 was a 15-horsepower, backwards-running tricycle design with the four-cylinder motor mounted crosswise on a pedestal above the single rear driving wheel. The rear drive wheel and the engine assembly pivoted at the frame for steering. Cultivator shovels were mounted just behind the two front wheels, and the operator sat just above the cultivator and in front of the engine. It was clumsy, as well as ugly.

Farmer acceptance of the motor cultivator concept was poor across the board. Row-crop farmers clung to their horse cultivators for such a critical task, and they resisted buying another machine that could only cultivate.

1918 International Motor Cultivator

In the ugliest and clumsiest farm machine ever made was an idea that eventually revolutionized row-crop farming. The 1918 IH Motor Cultivator was designed to replace three horses used to cultivate two-row crops, such as corn and cotton. From its precariously placed engine down to its narrow open drive wheel, the IHC cultivating tractor was an accident looking for a place to happen. Even with its two heavy cast front wheels, a slope in the field or a quick turn could topple it. Bill Splinter at the University of Nebraska, Lincoln, drives the machine.

Left: **1920s Farmall advertisement**

Below: **1927 Farmall Regular**

This 1927 Farmall is equipped with the front-mounted two-row cultivator that was made for the machine. The IHC tractors were painted gray until 1936, when red was used. The Farmall's 221-cubic-inch engine had four cylinders and rated at 20 horsepower at 2,100 rpm. Introduced in 1923, the Farmall was an all-purpose machine that could cultivate, as well as plow, disk, or power equipment from its belt pulley or PTO. The Farmall set the pattern for all row-crop tractors to follow, and it put IHC back on top as the number one U.S. tractor maker. Bill Montgomery of Columbia, Missouri, restored this historic machine.

1927 Farmall Regular
Wide wheel settings on the Farmall straddled two rows of corn with the tractor's tricycle front running between them. Extra crop clearance for "laying by" tall corn was provided by the drop gearboxes hanging from the axle. The wide drawbar was removed when the cultivator was installed.

IHC engineers began to realize there was potential for a machine that could do *all* of the power work on the farm—a real general-purpose tractor. Unsold motor cultivators—and IH had some—became engineering test vehicles as carriers for all sorts of mounted equipment. Engineering work finally evolved into a front-running tricycle-configured tractor to which a two-row cultivator could be mounted and then effectively steered from the rear operator's seat.

IHC engineers started calling it the Farmall before it was officially named or the trademark even registered. Twenty-two prototypes were put together in 1923 and sent to the field for testing. After constant testing, tweaking, and retesting of the design, 205 Farmalls were hand-assembled in 1924 and put on the market for field testing. They were well accepted.

Farmall production began in earnest in 1925; more than 800 were made. The figure rose to 4,418 in 1926, 9,501 in 1927, and

1934 McCormick-Deering W-12
The McCormick-Deering W-12 was a small standard-tread tractor of the same power as the row-crop F-12. Its 3x4-inch bore-and-stroke four-cylinder engine could burn either kerosene or gasoline and produced enough power to pull one big plow or two very small ones. It weighed about 2,900 pounds on steel wheels. The model was also available as an 0-12 orchard, an I-12 industrial, and as a Fairway 12 turf tractor with wide drive wheels. IHC collector James Gall of Reserve, Kansas, restored this 1934 tractor.

1936 Farmall F-12
An adjustable wide-front axle and steel lugs equip this 1936 Farmall F-12. Its 12 drawbar and 16 belt horsepower limited it to smaller jobs, but with its front-mounted two-row cultivator it could turn out productive field work. Many F-12s were equipped with mounted sickle-bar mowers, used to mow hay or clip small-grain stubble and pastures. James Gall of Reserve, Kansas, restored and painted this 1936 red F-12.

1936 McCormick-Deering O-12

The 1936 McCormick-Deering O-12 was a small orchard tractor for handling light jobs amidst the trees. Shielding over the tractor's rear tires protected tender blooms, buds, and fruit. Full orchard fenders helped keep the tractor out of trouble with the tree limbs, but the operator had to fend off limbs for himself. Paul Ganzel of Toledo, Ohio, restored this nifty little tractor for his IHC tractor collection.

1939 Farmall F-14
The smallest Farmall grew a little bit in 1939 when it became the Model F-14. With an increase in engine speed to 1,650, the F-14 tested at 14.84 drawbar horsepower and 17.44 belt horsepower and became a small two-plow tractor. A distinguishing feature of the F-14 was the elevated steering wheel mount. John S. Bossler, an IHC collector from Highland, Illinois, restored this beautiful 1939 F-14 with adjustable wide-front axle.

24,898 in 1928. By 1929, 35,320 Farmalls came off company lines. Farmall production peaked in 1930 at 42,092 tractors. Total production for the Farmall "Regular" was close to 126,000 from 1924 through 1932. The peak production years coincided with Ford moving Fordson production out of the United States to Ireland. IHC regained its lead in tractor sales on the strength of a new machine that firmly established row-crop tractors as the dominant farm tractor type. The popular models 10/20 and 15/30 standard-type plow tractors also aided the IHC first-place finish in tractor sales in 1928.

The 1924 two-plow Farmall was available with a front-mounted two-row cultivator, a rear-mounted mower, mounted middle breaker, and other attachments true to the "Farm-all" concept. The engine was an IHC four-cylinder 3.75x5-inch design running at 1,200 rpm. International didn't rate the Farmall's power initially, fearing competition with its own machines, specifically its 10/20 gear-drive tractor. When tested at Nebraska in 1925, the Farmall showed 9.35 drawbar horsepower and 18.3 belt horsepower—right in line with the 10/20 gear-drive standard tractor.

1939 McCormick-Deering W-40
Debuting in 1934, the big McCormick-Deering W-40 brought six-cylinder power, and eventually four-cylinder diesel power, to IHC standard-tread farm tractors. The 3.75x4.5-inch engine burned gasoline or distillate and cranked out 35.22 drawbar horsepower and 49.76 belt horsepower during its Nebraska tests. The WD-40 diesel was the first diesel-powered wheel tractor made in the United States. It had about the same horsepower as the six-cylinder spark-ignition tractor. Powell Smith of Shelbyville, Tennessee, bought this 1939 W-40 in 1977 and restored it.

Unique to the Farmall at its introduction was its high-standing, crop-clearing tricycle configuration. Its patented cultivator guidance was steering-connected to shift cultivator gangs quickly, and cable-actuated steering brakes facilitated short turns into the next row-pair at the end of the field. Belt pulley and PTO added to the overall utility of the Farmall. IHC was really onto something: a practical general-purpose tractor. Other manufacturers soon hustled to make their own row-crop or general-purpose machine.

A big brother to the original regular Farmall, the F-30 Farmall came out in late 1931. It was a larger tractor with the Farmall row-crop features. Its power was upgraded to three-plow size with a four-cylinder 4.25x5-inch engine operating at 1,150 rpm. The F-30 weighed close to 5,300 pounds and could turn around in an 18-foot circle. By 1936, it was available on factory-installed pneumatic rubber tires and with a new high-speed gear for road transit. The F-30 was built from 1931 to 1939 with nearly 29,000 tractors made. W-30 standard-tread versions sold another 32,000 units. Industrial I-30s and the W-30s were made from 1932 to 1939. Beginning in late 1936, IHC tractors were painted the now-familiar Farmall red.

The original Farmall Regular was replaced with the new Model F-20 in 1932. Much like the original Farmall, the F-20 was

Opposite, top and bottom: **1939 McCormick-Deering W-30**
Based on the engine from the F-30 Farmall, the small but mighty McCormick-Deering W-30 was developed in 1932 to replace the retiring gear-drive 15/30. As tested at Nebraska, the W-30 put out 19.69 drawbar horsepower and 31 belt horsepower. Its three-speed transmission was updated to a four-speed in rubber-equipped tractors. The W-30 was a full three-plow tractor. It had a belt pulley for powering threshing machines, corn shellers, feed grinders, and wood saws. Paul Ganzel of Toledo, Ohio, restored this fine 1939 W-30 on rubber for his growing IHC collection.

slightly larger and had 10 percent more power. Nebraska tests showed the F-20 could produce 24.13 belt horsepower and 16.12 drawbar horsepower. The F-20's four-speed transmission gave it operating speeds of 2.25, 2.75, 3.25, and 3.75 miles per hour. In the mid-1930s, it became available on rubber tires. A record number of nearly 149,000 F-20s had been made when production was ended in 1939. Options for the later F-20s included adjustable wide fronts, wheel weights, and an electric starter and lights.

A BABY FARMALL

A small Farmall, the Model F-12, was introduced in 1932. Rated as a one- to two-plow tractor, it took the row-crop tractor design to its final form. Long rear axles with a machined-in keyway permitted the rear wheels to be adjusted to any width by first loosening the hub bolts, then sliding the wheels on the axles to the desired setting. Its four-cylinder 3x4-inch bore-and-stroke engine had an operating speed of 1,400 rpm. Nebraska tests showed the

1939 Farmall F-20
All of the options available for the Farmall Model F-20 are included on Tom Hill's 1939 tractor. But it took Hill seven years to find all of the parts for this restoration. IHC introduced the Model F-20 in 1932 as a replacement for the original (regular) Farmall of 1923. Power was boosted about 10 percent to 16.12 drawbar horsepower and 24.13 belt horsepower. This tractor, among the last of the F Series, has an adjustable wide-front end, lights and electric starter, wheel weights, rubber tires, and fenders. Hill collects IHC tractors and farms with his family near Piqua, Ohio.

gasoline-burning F-12 produced 12.31 drawbar horsepower and 16.2 belt horsepower.

Equipment made for the F-12 included two-row mounted cultivators, rear-mounted PTO-driven mowers, mounted planters, plows, and other implements. Quick-Attach features speeded changeover between implements or replacing the drawbar. Long toggle bolts flipped over the tractor axle housing, dropped into slots on the mounting clamps, and secured the rear implement attachments. A long single-purpose wrench with an offset crank handle speeded up the job of tightening the clamps.

Rubber tires were made available as a factory-installed option early in the F-12's run, and a special road gear was available for rubber-tired models. The regular three-speed transmission provided ground speeds of 2.25, 3, and 3.75 miles per hour. The F-12 was another IHC success. A total of 123,407 were made from 1932 to 1938. A standard version, the W-12, was available from 1934 to

1942 Farmall H
Rubber shortages during World War II caused the few tractors built at the time to be equipped with steel wheels. This 1942 Farmall Model H wartime tractor is an example. It was also built without starter and lights. The Model H Farmall, a third-generation row-crop tractor, used large rear wheels mounted on a solid one-piece axle to provide under-axle clearance for cultivated crops. The heavy drop gearboxes introduced on the first Farmall were thus eliminated. David Morrison, of Port Deposit, Maryland, restored this wartime machine.

World War II shortages
Rubber wasn't the only thing in short supply during the war. Farm workers also were depleted, as many young men joined the service and were fighting overseas. As a result, women had even a stronger presence in American fields.

1938. Only 3,617 W-12s were made. The O-12 orchard tractor was available from 1935 to 1938 and sold nearly 2,400 units. Another version, the Fairway-12, with wide rear tires, was built for golf course maintenance and other turf operations.

The F-12 was replaced in 1938 with the F-14. It was much like the earlier tractor, but IHC sped up the engine to 1,650 rpm compared with 1,400 on the F-12. On rubber tires, the F-14 tested out at 14.84 drawbar horsepower and 17.44 belt horsepower at Nebraska. Standard, fairway, and orchard versions of the Model 14 were available during 1938 and 1939. Total F-14 production was 27,396, with another 1,900 tractors made in the standard-tread versions.

MAKE WAY FOR DIESELS AND CRAWLERS

IHC additions to its big standard tractor line in 1934 were the six-cylinder W-40 and WD-40. The WD-40 diesel was the first U.S. wheel-type tractor to be diesel powered. The W-40 gas engine was the first six-cylinder engine IHC used in a farm tractor. The W-40 had a six-cylinder 3.75x4.5-inch engine and burned kerosene or distillate.

Nebraska tests rated the W-40 at 35.22 drawbar horsepower and 49.76 belt horsepower. The diesel WD-40 did about as well with 37.21 drawbar horsepower and 48.79 horsepower on the belt. Big farmers now had a big tractor. Both were available on rubber or steel. The W-40 was the more popular unit—6,454 were made,

1952 McCormick-Deering O-4
International Harvester reserved the Farmall name just for its line of row-crop tractors. Other IHC tractors had the McCormick-Deering label, such as this 1952 O-4 orchard tractor. This is the equivalent of the Farmall Model H in orchard dress. It is a two-plow machine with metal shielding to prevent tree damage when working in orchards and groves. Veteran tractor collector Verlan Heberer of Belleville, Illinois, includes this O-4 in his orchard tractor group.

1950 McCormick-Deering W-4
Primary tillage work preparing the soil for corn and soybean crops was the lifetime work of this 1950 McCormick-Deering W-4 Standard. Walton DeCook of New Sharon, Iowa, cherishes his little workhorse from the past.

while only 3,370 of the diesels came out between 1934 and 1940. Industrial versions of the gas and diesel W-40s were made from 1936 to 1940.

Crawler or track-type tractors entered the IHC line in 1928 with a small machine based on the 10/20 tractor. The 10/20 TracTracTor was built in only 1,504 units until it was changed to the newer T-20 in 1931. The T-20 TracTracTor weighed in at about 6,725 pounds and ran the 10/20 engine at a faster clip of 1,250 rpm. It was built between 1931 and 1939 and was the most

popular of the IHC farm-size crawlers with 15,198 units produced. Larger crawlers, the TA-40, TK-40, and TD-40 (gas, kerosene, and diesel models), were produced from 1932 to 1934 in numbers totaling 1,312. They were then replaced with the new T-40 and TD-40 models with more than 7,600 of them built between 1934 and 1939. The 40 Series crawlers used the same engines as the W-40s. Nebraska tests showed them pulling more than 40 drawbar horsepower. More than 7,060 of those tractors were made by the time production ended.

1940 Farmall M and 1953 International M-21 Till-Planter
The Farmall Model M row-crop tractor was the IHC power leader of its category. The "muscular" M was a full three-plow tractor with 39.23 maximum belt horsepower on gasoline. The Model M and Model H Farmalls shared a common wheelbase and attaching points, so mounted implements would fit either tractor. This 1940 M was restored by the Agricultural Engineering Department at Purdue University, West Lafayette, Indiana. Mounted on the tractor is a 1953 IH M-21 Till-Planter. It was an early reduced-tillage system for corn culture.

More than 5,580 Model T-35 and TD-35 TracTracTors were produced from 1937 to 1939. The T-35 was a mid-size crawler and could pull 35 horsepower on the drawbar. The small farm crawlers built before 1940 formed the base for IHC's later entry into the industrial crawler market.

A BRAND NEW LOOK

The sharp corners and square fronts on IHC's tractors were smoothed and rounded in the late 1930s. Famed industrial designer Raymond Loewy began his work with the company by first redesigning the IH trademark and then moved on to improve the tractors' looks and operations. The big, new, 80-horsepower TD-18 crawler tractor introduced in 1938 was the first unit out that showed the Loewy touch. The distinctive rounded radiator grille with its horizontal slits soon dressed up all of the crawler line, including the smaller T-6, TD-6, T-9, TD-9, T-14, and TD-14 models.

The big news from IH in 1939 was its spanking new line of Farmall tractors. They, too, carried the Loewy styling. The largest row-crop design, the Farmall M, eventually became one of the all-time great classic American tractors. Produced on the same

1941 Farmall MD
One of the first practical row-crop diesels, the 1941 Farmall Model MD shared the same bore and stroke as the gas and distillate M engine. It was started on gasoline and switched over to high-compression diesel and diesel fuel injection as the engine warmed up. The MD produced about 37 belt horsepower. IHC pioneered the use of diesel engines in wheel tractors with the introduction of the WD-40. IH also offered diesel engines in some of its crawler tractors. Contractor Alan Smith of McHenry, Illinois, bought this MD locally and restored it to its past glory.

wheelbase to maximize implement interchangeability was the smaller-engined Farmall H.

The Model M and H tractors no longer had geared drop boxes in the rear. Large rims and tires raised the straight axles for the needed row-crop clearance. Their single-keyed rear axles allowed tread adjustment by sliding the wheel hubs to the desired width setting. The three-plow Farmall M had optional electric starting and lights, belt pulley and PTO, and hydraulic Lift-All. Rubber tires were standard, but it could be ordered on steel wheels for less money. The diesel version, the MD, first appeared in 1941. Like

the earlier IHC diesel engines, the MD started on gasoline and switched to diesel once it was running. The gasoline M produced 25.83 drawbar horsepower and 33.35 belt horsepower in its Nebraska tests. The Model M kerosene tractor results were somewhat lower. The M engine had a 3.875x5.25-inch bore and stroke and was designed to operate at 1,450 rpm. More than 297,000 Model M Farmalls, including its specialized versions, were manufactured from 1939 to 1952. The McCormick-Deering standard version in the M size was the McCormick-Deering W-6. It was also made as the OS-6 and O-6 in orchard modifications.

1940 McCormick-Deering O-6

Specially designed to work in orchards without harm to
the trees or machine, this 1940 McCormick-Deering
Model O-6 orchard tractor was equivalent in power to the
three-plow row-crop Model M Farmall. Dan Walters of
Carey, Ohio, invested his time, skill, and money in
careful restoration of the nearly seventy-year-old tractor.
The spring apple blossoms in the background are part of
the landscape at Toledo Ohio's McQueen Orchards.

1941 Farmall A

The Farmall A featured "Culti-Vision" for accurately following the one row being cultivated. The A's driver sat to the right, the engine and drivetrain were positioned to the left, and the crop row passed under the center of the rig. With its small rear tires, the Farmall A used dropped gearboxes from the axle housings to provide under-axle crop clearance. The A's 3x4-inch engine created 16 drawbar horsepower at 1,400 rpm. Thalua and Alton Garner of Levelland, Texas, found this Model A in New Mexico and put a lot of time and care in restoring the 1941 tractor.

The two-plow Farmall H had the same equipment options as the larger M. Its 3.625x4.25-inch engine ran at 1,650 rpm. In Nebraska tests, the gasoline-powered H put out 19.13 horsepower at the drawbar and 23.72 horsepower on the belt. The kerosene version showed a couple of horsepower less in each category. Between 1939 and 1952, more than 391,000 two-plow Hs came out, surpassing the M Farmall production and sales by nearly 100,000 tractors. Despite the iconic stature of the M, the Farmall Model H was the most popular tractor of its era.

The standard-tread counterparts to the Farmall H were the McCormick-Deering W-4, OS-4, and O-4 models. The OS-4 had the exhaust and air cleaner mounted under the hood to reduce its

height for orchard work. The O-4 had complete shielding of the rear tires and the operator platform to protect orchard and grove trees. Orchard models were available with many of the same options as available on the Farmalls.

Two new small tractors, the Model A and Model B, had the same size engine, but different appeals. The Model A was a small machine with wide-set, adjustable-tread front wheels. Its engine and drivetrain were offset to the left of the frame. The operator sat at the right of the engine for a better view of the cultivator. IH called the feature Culti-Vision. The Model B had a centered engine and drivetrain, but its operator sat on the right rear axle for its Culti-Vision feature. The B came with a

1946 Farmall B

The Farmall B was a two-row tricycle-configured tractor with some of the Cultivision features of the Model A. Although the engine and drivetrain were centered on the rear axle, the operator sat to the right. The Model B had its two-row cultivator positioned so the right row passed directly under the operator. The B was powered with the same 3x4-inch engine used on the one-row Model A and was considered a light two-plow machine. Bill Scott of Hawarden, Iowa, brought his 1946 Model B back to like-new condition.

1949 Farmall Cub
Smallest of the Farmalls was the little Cub, introduced in 1947. The IHC four-cylinder 59.5-cubic-inch Cub engine put out 8.30 belt horsepower at 1,600 rpm in initial tests. Mounted implements, including belly-mounted mowers, made the Cub a valuable estate-care tractor. It received many power tweaks in its thirty-two-year production life from 1947 until 1979. During that time, 152,997 Cubs were made. Lawrance N. Shaw, a retired agricultural engineer from Gainesville, Florida, used his shiny 1949 Cub to pull an automatic vegetable transplanter he developed.

tricycle front and had enough rear tread width to cultivate two rows instead of one.

The 3x4-inch bore-and-stroke engines in the Model A and B Farmalls showed about 13 drawbar horsepower and 16 belt horsepower when gasoline-fueled. The distillate tests were nearly 2 horsepower less. Engines in both models ran at 1,400 rpm.

Combined, the As and Bs accounted for more than 218,000 unit sales during their production from 1939 through 1947.

IHC's big wheel-type farm tractor of the 1940s was the McCormick-Deering W-9 standard. It was four-plow size with a big 50-horsepower pull shown at the drawbar. It was available as a diesel WD-9 and in rice field versions as WR-9 and WRD-9. It

1951 Farmall Super C
The new Farmall Model C of 1948 replaced the previous Model B. The C was a light two-plow, two-row tricycle row-crop tractor that placed the operator high on the machine's centerline. Live hydraulic Touch-Control raised, lowered, and depth-adjusted its front-mounted cultivator or other implements. Its 3x4-inch four-cylinder IHC engine provided 22.18 belt horsepower. The Farmall C became the Super C in 1951. Kelley Shoemaker of California, Missouri, owns this sparkling 1951 Super C.

was marketed between 1940 and 1953. More than 67,000 W-9s of all types were made during that timespan.

World War II shifted much farm equipment production into defense items. Steel wheels temporarily replaced rubber tires until synthetic tires became available. New model introductions were rare. Tractor production nearly ceased. Postwar, materials were scarce, so it took time for the farm equipment industry to supply the pent up demand.

CUBS AND CS

Two different-sized Farmalls replaced the Model B in 1947: the Farmall Cub and the Farmall Model C. The Cub was a tiny Farmall designed for one-row cultivation and estate manicure. It could pull about 9 drawbar horsepower with its 2.625x2.75-inch engine turning at 1,600 rpm. It featured the offset engine and drivetrain for improved Culti-Vision like the A and B Farmalls. More than 185,000 Cubs were made from 1947 to 1954.

Farmall Cub advertisement

Above and opposite: **1953 Farmall Super H**

The Farmall H became the Super H in 1953, sporting a stronger pull of 33.4 drawbar horsepower as IHC ratcheted up the power of its tractor line. Made only in 1953 and 1954, the Farmall Super H sold well—better than the larger Farmall M. In this photo, Max Armstrong, a WGN radio and TV farm broadcaster, has parked his splendid 1953 Farmall H on the Chicago River bank, near where Cyrus McCormick's first reaper plant once stood. Armstrong's Farmall H came from his home farm near Owensville, Indiana.

The Model C was a modernized, light, tricycle row-crop with improved Touch-Control hydraulics to ease the operation of its attached implements. Unlike its Model B predecessor, the operator sat centered behind the engine and drivetrain. It had single-piece keyed axles for rear-wheel tread adjustability. Visibility to the two-row attachments was enhanced by narrowed frame components ahead and below the operator. The handy little C was built in numbers approaching 80,000 during its manufacture from 1948 to 1951.

MORE POWER FOR THE FIFTIES

Farmers were clamoring for increased horsepower to more efficiently power their larger farms as the 1950s dawned. Outside suppliers offered kits to "soup-up" tractor power, but those only temporarily met farmers' increasing power needs. IHC joined the horsepower race in 1952 with its Super M Farmall with Torque Amplifier, or Super MTA model. The torque amplifier was also added to the Farmall Super MD diesel and the Super W-6 standard-tread machine.

The TA provided two speeds in each transmission gear, giving the Super M Farmall and the other TA-equipped tractors the equivalent of ten forward and two reverse speeds. Lever-operated, the TA could be engaged "on the go" to reduce speed by about one-third but increase pulling power by 48 percent. The operator could also shift back up to the previous speed without stopping. IHC also offered live PTO on its new tractors in 1954. PTO-driven implements could now be run independently of tractor ground travel. Hydraulic implement control was also increasingly available. The first hydraulics were available as remote cylinders on implements, then they were coupled with lift arms on the tractor.

First of the Super series was the Super C Farmall, produced from 1951 to 1954. Its four-cylinder mill was a 3.125x4-inch engine that furnished 20.72 drawbar horsepower and 23.67 belt horsepower in Nebraska tests—a couple more horses than the older Model C. In 1953, a two-point Fast-Hitch was made available on the Super C. It was designed to compete with the three-point hitch first offered on the Ford-Ferguson tractor. The Super C also

offered optional adjustable wide front tread. More than 98,000 Super C Farmalls were made before it became the Farmall 200 in 1954.

The A became the Super A in 1952, while the B was dropped in 1947. The Model H became the Super H in 1953 with an improvement in horsepower output to 30.69 drawbar horsepower and 33.4 belt horsepower. It was replaced in 1954 by the Farmall three-plow 300 that offered IHC's Fast-Hitch and Torque Amplifier. The Super M's successor was the Farmall 400, built from 1954 to 1956. It was available with Fast-Hitch and Hydro-Touch hydraulics in either gas or diesel. The W-400 and WD-400 were its standard-tread versions. The 400's 4x5.5-inch gas engine produced 45.34 drawbar horsepower and 50.78 belt horsepower in Nebraska tests.

1954 Farmall Super A
With its 1939 Raymond Loewy styling intact, the Model A became the Farmall Super A, like this 1954 model. The one-row cultivating tractor had Touch-Control hydraulics, as well as electric lights and a starter. Cultivision features first seen on the 1939 Farmall A and B models were retained on the 1954 Super A. Horsepower in the Super A was bumped up to 20.7 on the drawbar, or about 4 more than the previous Model A put out. Jerry Baltisberger of Grinnell, Iowa, is proud of this beautiful Super A that he restored.

International 300 utility tractors became available starting in 1955. Their low profile and standard tread made them ideal loader tractors. The utility series became especially popular on livestock farms, where they became the chore tractor and could be pressed into tillage and planting work when the season demanded.

The Super A became the 100 in 1955. The 100 Series tractors had a restyled radiator grille with central vertical slots added to the widened horizontal slots in the grille. Model numbering and the Farmall or International name was in chromed letters and numerals. The IH logo, also in chrome, was prominently displayed on the top front of the radiator grille.

In 1956, the Farmall 130, 230, 350, and 450 replaced the even-numbered models of the year before. A cream color was used in the paint scheme as a backing for the Farmall trademark on the hoods and as a contrast on the radiator grille. The 130 replaced the 100, the 230 replaced the 200, the 350 the 300, and the 450 the Model 400. Horsepower was increased on all of the tractors. The International 330 utility came along in 1957.

Major product improvements were included in the 1958 IH model lineup. The Farmall 140, 240, 340, 460, and 560 featured all new styling. The radiator grille was squared off with bold horizontal bars in the grille. The cream accent now swept back from the

1954 Farmall MTA
The ultimate Farmall M of its day was this 1954 Super M with Torque Amplifier. The lever-shifted torque amplifier gave the tractor on-the-go shifting. Forward speed dropped 32 percent and tractor pull increased by 48 percent when shifted down to pull through tough spots. The Super MTA produced about 42 drawbar power and had live PTO. IHC fan Gary VanVark of Pella, Iowa, restored this 1954 Farmall Super MTA with wide-front adjustable axle.

1950s Farmall advertisement

IH Torque Amplifier advertisement
International Harvester first introduced its Torque Amplifier (TA) on the Super MTA model, but soon offered it on other models to boost these machines' performance.

grille and shroud to the sides of the hood. The Farmall 460 and 560 models came with six-cylinder engines in gasoline or diesel. They were the first row-crop tractors built by IH to have six-cylinder engines.

The 460 and 560 were popular because of their powerful six-cylinder engines, available in gasoline or diesel. The hydraulic pump (by now a serious part of tractor systems) was moved inside the transmission case from the previous engine block location. A crankcase oil cooler was added to help dispel heat from the hard-working engines. Steering was reconfigured from the over-the-top position of the early Farmall to the shaft placed along the engine side rail. There it connected to a worm drive inside the front bolster where steering was hydraulically aided. Neat instrument consoles were positioned near the steering wheel. Seat backrests and deep-cushioned seats addressed operator comfort.

In 1960, IH continued with many added improvements to the Farmall line. The new 404 and 504 models boasted the first American-designed three-point hitch with draft sensing. In 1959 industry standards for three-point hitches had been adopted to enhance the interchangeability of implements between different tractor makes. IH also took its crankcase cooler another step forward and used it to cool hydraulic fluid. A dry-type air cleaner was adopted for use in the new models, and IH adopted hydrostatic power steering in the 504.

Although they didn't completely replace the tricycle-type tractor, more row-crop Farmalls were available with adjustable

1956 Farmall 300

Bought new in 1957, this 1956 Farmall Model 300 served for ten years as the main tractor for farmer Henry L. Nunnikhoven of Pella, Iowa. He has since restored it to like-new condition. The Model 300, with chrome nameplates, replaced the Super H model in the Farmall line. The 27–drawbar horsepower three-plow tractor offered the Torque Amplifier, two-point hitch, and full hydraulics. Electric lights, starter, power steering, and live PTO were standard equipment. The Farmall Model 300 was built from 1954 to 1956.

1954 Farmall 400

Farmall 400 owner Don Rimathe, who farms near Huxley, Iowa, was only nine when he learned to drive this big IH row-crop owned by his father, Ray. Don has kept the tractor and carefully reconditioned it to like-new shape. This 1954 machine replaced the Super M and Super MTA Farmall in 1954 and was made until 1956. At 45.34 drawbar and 50.78 belt horsepower, it was a real producer. Its two-point Fast-Hitch, Hydra-Touch hydraulics, live PTO, and Torque-Amplifier gave the 400 tools to facilitate its field work. The Farmall 400 was available in gas, diesel, and LPG versions. It was also available as a McCormick-Deering Model W-400 standard-tread machine.

1957 Farmall 350

A new Harvester White trim color marks the grille and side panel of this 1957 Farmall 350. Announced in 1956, the new model series added Traction-Control, Fast-Hitch, and spin-out tread adjustment. A diesel version of the 350 was available with a four-cylinder Continental diesel of 3.75x4.375 bore and stroke that tested at 36.26 drawbar horsepower and 38.65 belt horsepower. Don Rimathe of Huxley, Iowa, parked his 350 gasoline Farmall next to a field of his tall Iowa corn.

1957 Farmall 450
Four-plow, four-row power was the forte of the Farmall Model 450, introduced in 1956. The diesel model demonstrated 45.17 drawbar horsepower and 48.78 belt horsepower at Nebraska in 1956. Gasoline and LPG tests were similar, with the gasoline version producing 55.28 belt horsepower. William J. Scott of Hawarden, Iowa, restored this 1957 Farmall 450 jewel.

1957 Farmall 230
The smallest two-row Farmall in the 1956 model lineup was the 230. With a family history tracing back to the Model C of 1948, the Model 230 had grown in power to nearly 25 drawbar horsepower and could pull two 14-inch plows. Don Rimathe of Huxley, Iowa, restored this 1957 Farmall Model 230. Front weights counterbalance implements mounted behind in its two-point Fast-Hitch.

wide-front tread. With industry adoption of the three-point hitch in 1959, more implements were mounted on the rear of the tractor as integral hitches. That included rear-mounted cultivators, so there was no longer the pressing need to mount cultivators in front of the tractor beside the quick-turning tricycle front.

RAPID CHANGES TOWARD THE FUTURE

Bold new advances in engineering, power, design, and manufacturing marked the next two decades for IHC. Diesel engines gained general acceptance, and they were increasingly tweaked to produce more horsepower and efficiency. Turbocharging crammed more oxygen into the new engines. Intercooling was added to make the turbocharged air even denser, and transmissions became more sophisticated with the need to transmit more power. First, mechanical or hydraulic front-wheel assist and then

four-wheel-drive was increasingly used as available horsepower passed the 150 mark and headed toward 200. Factory-built cabs were added to the bigger machines.

International Harvester announced what it called the "world's most powerful four-wheel-drive agricultural tractor" in 1961. The Model 4300 was a ten-plow 300-horsepower diesel tractor with a Category IV three-point Fast-Hitch that could haul and control the big-mounted plow. The 4300 was impressive, but way ahead of the practical needs of the IH customer base. It was an obvious competitive answer to John Deere's 215-horsepower four-wheel-drive Model 8010, introduced in 1959. Both companies soon backed off to lower-horsepower ratings on four-wheel-drives in their later model introductions.

IH offered four-wheel-drive again in 1965 with its Model 4100 diesel. Its 429-cubic-inch six-cylinder 4.5x4.5 bore-and-stroke

1959 Farmall 560 Diesel

Six-cylinder power and all-new styling were features of the 1958 Farmall model lineup. The 1959 Farmall Model 560 six-cylinder diesel had the horsepower farmers were seeking. It was a five-plow machine powered by a 281-cubic-inch diesel that tested 54.88 drawbar horsepower and 59.48 belt horsepower at Nebraska. The diesel was direct-starting with the use of glow plugs. Gasoline and LPG sixes were also available. Farm implement maker Jon E. Kinzenbaw of Williamsburg, Iowa, had this 1959 Farmall 560 restored for his growing collection of IHC tractors. *Above:* The 560 Farmall's hydraulics were controlled with these Hydra-Touch levers.

diesel engine was turbocharged and proved 110.82 drawbar horse-power and 116.15 PTO horsepower in Nebraska tests. The 4100 four-wheel-drive was the first IH farm tractor to be turbocharged. It featured an optional factory-installed cab with heater and air conditioning, as did the smaller Farmall 1206 diesel introduced that year.

The Model 1206 Farmall, introduced in 1965, was the first IH two-wheel farm tractor tested with more than 100 PTO horse-power. It also had the distinction of being the first Farmall tractor with a factory-equipped turbocharger. The Farmall 1206 model was driven by a 361-cubic-inch six-cylinder diesel engine rated at 2,400 rpm. Its Nebraska review gave it 99.16 drawbar horsepower and 112.64 PTO horsepower.

Infinitely variable speeds were available in the first hydrostatic drive tractors from IH in 1967. The tractors used a variable displacement hydraulic pump and motor in their high-low transmissions to offer an infinite number of speeds from zero to 20.5 miles per hour. The Farmall 656 Hydro came as either a gaso-line- or diesel-powered unit, which tested at 65.80 and 66.06 PTO horsepower, respectively. Some power was lost to the hydraulic transmission, and both versions tested at about 50 drawbar horsepower. Another somewhat smaller Farmall model, the 544 Hydro, was put in the IH line in 1968. It had a 53.87-PTO horse-power gas and 55.52-horsepower diesel engine. Also available as the Farmall 544, the tractor could be ordered with a conventional gear transmission. The 112.45-PTO horsepower International 1026 Hydro diesel was tested in 1969. It transferred 83.94 horse-power to the drawbar via the hydrostatic drive. Its power came from a 407-cubic-inch six-cylinder turbocharged IH diesel engine.

International Harvester manufactured its milestone five mil-lionth tractor in February 1974 and showed the specially-marked International Model 1066 Turbo diesel at fairs and farm shows that year. The company had earlier discontinued the use of the Farmall brand name to mark tractors as row-crop machines.

Tractor innovation continued in the 1970s, and in 1979 International's 2+2 articulated four-wheel-drive row-crop trac-tor was introduced. It was available as the 130-horsepower Model 3388 or the 150-horsepower Model 3588. The tractor's unique design positioned the engine in front of the front axle on the forward segment of the hinged frame. The operator sat in a cab over the rear wheels. The 2+2 promise was that of a nimble row-crop machine with the power of a highly productive large horsepower tractor.

THE FATE OF THE FAMOUS FARMALLS

The future of the International Harvester Company took a big hit during the farm crisis of 1980. Kicked off by a grain export embargo, commodity prices dropped, interest rates peaked above 20 percent, and land prices plunged, causing farmers to stop buying equipment. On top of that, IHC workers went on a six-month strike. By the end of fiscal 1980, IHC losses reached $397 million. By 1981, IHC's 150th anniversary year, losses were at $393 million. The next year, 1982, proved disastrous with losses mounting to $1.638 billion. While the company curtailed its 1983 losses to $485 million, the damage had been done. IHC couldn't turn itself around.

In late 1984, Tenneco Inc. bought the agricultural equipment division of IHC and brought it into its J. I. Case Company sub-sidiary. The merged name was changed to Case-International the following year, and International Harvester and its famous Farmalls were no more.

Today, International Harvester lives on within the Case-IH group of CNH Global N.V., headquartered in Amsterdam. CNH (based on the initials of Case New Holland) is 91 percent owned by the Fiat group of Italy. It is the second-ranking producer of farm equipment today—behind Deere & Company of Moline, Illinois, which still holds a commanding lead as the premier maker of trac-tors and farm equipment.

Case-IH farm equipment is still made and sold under that brand. Its largest and most impressive product today is the Case-IH Steiger Quadtrac tractor. The Model STX530's rubber-belt crawler tracks are driven by a 530-horsepower, 15-liter diesel engine. Like the Steiger design of years past, it is of articulated design and bends in the center as it turns. Case-IH also offers a full range of wheel tractors, with power rankings from 19 horsepower up to 530 horsepower in a four-wheel-drive unit.

The famous Farmall trademark, once used by IH to distin-guish its row-crop tractors, recently has been revived by Case-IH to brand its DX Series of subcompact and utility tractors. The company made the announcement in 2004. The series of handy tractors, now being sold as Farmalls, feature options such as four-wheel-drive, hydrostatic transmissions, and three- and four-cylinder diesel engines from 30 to 55 gross horsepower.

1967 Farmall 1206 Turbo Diesel
The first Farmall with a factory turbocharged diesel engine was the 1965 Farmall 1206, which also produced more than 100 horsepower. The 1206's 361-cubic-inch six-cylinder turbocharged engine developed 99.16 drawbar horsepower and 112.99 PTO horsepower when tested at Nebraska in 1965. A factory-installed cab with heater and air conditioning was optional. IHC also introduced the four-wheel-drive International 4100 in 1966. Greg Kokemiller of Madrid, Iowa, had this powerful 1967 Farmall restored.

Chapter Eight
Massey-Ferguson

1950 Massey-Harris 44 Vineyard
Built low and narrow to slip through vineyards, the Massey-Harris Model 44 Vineyard was a rare machine—only thirty were made. The Model 44 tractor was built between 1947 and 1955 either with a four- or six-cylinder engine. This 1950 example is owned by tractor collector extraordinaire Larry Maasdam of Clarion, Iowa. Its restoration was completed by Dick Carroll of Alta Vista, Kansas.

MASSEY-HARRIS

Massey-Ferguson, the shortened name of Massey-Harris-Ferguson, Ltd., formed in 1953 when Massey-Harris Company merged with Harry Ferguson, Inc. Ferguson was the inventive Irishman, who together with Henry Ford, brought out the 1939 Ford-Ferguson 9N tractor with its hydraulic draft-sensing three-point implement hitch.

The 1891 merger of the Massey Manufacturing Company of Toronto, Canada, and the A. Harris, Son & Company Ltd. of Brantford, Ontario, initially put together the Massey-Harris Company, Canada's largest agricultural equipment firm. The Massey firm has roots tracing back to 1847, when Daniel Massey started making simple farm implements. About ten years later, Alanson Harris started making similar tools. The two companies eventually grew to the point that they were numbers one and two in Canadian farm equipment sales. As Massey-Harris, they continued to add Canadian firms and in 1910 got into the U.S. market with the purchase of Johnston Harvester Company of Batavia, New York.

MASSEY GETS SERIOUS ABOUT TRACTORS

Massey-Harris sold about every other farm implement, but it needed a tractor to become a full-line company. The firm sold the three-wheel Bull tractor from Minneapolis for a short time, starting in 1917, but the Bull's rapid drop in popularity doomed that effort. The next tractor Massey-Harris offered was the Parrett, originating in Chicago, Illinois. Massey-Harris (Parrett) models No. 1, No. 2, and No. 3 were probably assembled in Weston, Ontario, between 1918 and 1923—the same time period the little gray Fordson was making inroads into all companies' tractor sales. When the Fordson was imported into Canada, it effectively killed Parrett sales there as well as it had in the United States.

In 1926, Massey-Harris began to negotiate marketing rights for the Wallis tractor line with its owner, the J. I. Case Plow Works of Racine, Wisconsin. The Plow Works was separate from the J. I. Case Threshing Machine Company of the same city. Beginning as early as 1902, plow works President Henry M. Wallis, J. I. Case's son-in-law, was working on a big gas tractor made in Cleveland, Ohio, called the Wallis Bear. The Wallis

Big Bull
Introduced in 1915, the Big Bull tractor was a three-wheeled machine with only its furrow wheel doing the pulling. The Big Bull could also power up its land wheel to help in poor traction conditions. Made in Minneapolis, Minnesota, by different companies, the Bull initially garnered favorable attention because it was rather lightweight and inexpensive. The Big Bull weighed 4,500 pounds and was powered with a two-cylinder opposed engine of 12/24-horsepower performance. In 1917, Massey-Harris sold the tractor briefly. Robert Newman of Slater, Missouri, found this rare machine and restored it.

Tractor Company merged into Case Plow Works in 1919, and the Wallis tractor developed into an early practical tractor that was relatively lightweight and powered by a four-cylinder engine mounted lengthwise on the frame.

THE WALLIS TRACTOR HERITAGE

From its start about 1902, the Wallis Bear was a 40/80 giant designed to do only the big jobs. It was followed with a smaller 20/50 tractor. But compared with the other gas tractors of its day, the Bear was relatively modern. The Bear started as a four-cylinder with the radiator-cooled engine mounted inline on the frame. The Bear's 1912 version was slimmed down to a mere 10.5 tons and was rated at 30 drawbar horsepower and 50 belt horsepower. Its 7.5x9-inch engine chugged along at 650 rpm. Only nine of the models are known to have been built.

Wallis cut the Bull's tractor size and horsepower in half for his next model: the 1913 Wallis Cub. It too had a four-cylinder vertical lengthwise-positioned engine. A boilerplate crankcase and transmission housing served as the Cub's frame. It was rated as a 26/44 machine from its 650-rpm engine that could burn gasoline, kerosene, or distillate. Like its Bear predecessors, the Cub featured one wheel in front with the belt pulley in the rear.

Big Bull
The Big Bull's drawbar for connecting implements is centered in the line of draft behind the big, single drive wheel.

In 1916 the Cub's size was reduced by half again in the Cub Jr., or Model J. It was rated a 13/25 tractor with a four-cylinder 4.5x5.75-inch bore-and-stroke engine. The Model J had an improved design with the boilerplate unit frame housing the engine crankcase, the transmission, and the differential. All drive components were fully enclosed—quite a feat for a 1916 machine. The belt pulley was squeezed into a position on the left side of the tractor between the wheel and the transmission housing.

Wallis based subsequent tractor designs on the successful Cub Jr. model, but dropped its single front wheel in favor of two wide-spaced front wheels with automotive-type steering on its Model K in 1919. The 1922 Wallis Model OK was tested at Nebraska in 1922 and turned out 18 drawbar horsepower and 28 belt horsepower from its 4.25x5.75-inch engine turning over at 1,000 rpm. By that date, the tractor weighed about 4,000 pounds. More engine speed and other tweaking turned the 1927 Wallis OK into a 20/30 tractor. Massey first sold it as the Wallis 20/30 "certified" under a marketing agreement with Wallis.

MASSEY BUYS A TRACTOR LINE

In 1928, after first selling the Wallis tractors in Canada and parts of the United States under a sales agreement with the J. I. Case Plow Works of Racine, Wisconsin, Massey-Harris bought the entire company. It acquired not only the Wallis tractor line, but also the company's line of plows and other implements as well. One thing it didn't need out of the deal was the J. I. Case Plow Works name. Massey-Harris subsequently sold to the J. I. Case Threshing Machine Company all of its acquired rights to the Case and J. I. Case names. Case then shortened its name to the J. I. Case Company.

In 1929, Massey-Harris brought out its upgraded Wallis 20/30 as its Wallis 26/41, or Model 25. It also introduced a smaller version, the 12/20. The 12/20 had a less powerful engine with 3.875x5.25-inch bore and stroke that drove a three-speed transmission. Its weight was down to 3,432 pounds.

In 1930, M-H announced its first general-purpose or row-crop tractor. Ahead of its time, it was an all-new four-wheel-drive

1910s Big Bull advertisement

machine. Four tread widths were available to match different row-crop row spacings. The front wheels turned to steer the tractor with power transmitted through universal joints. The rear-axle assembly pivoted at the rear differential housing to allow its four wheels to stay in ground contact on uneven terrain. A two-row cultivator was available for mounting on the front of the tractor. By 1936, an improved general-purpose tractor was available on rubber tires, but it didn't sell well.

1912 Wallis Bear
Forward or backward pressure on the Bear's steering wheel engaged the friction cone clutch power-assist steering for the tractor's front wheels. The driver needed all the help available when the big tractor was stopped.

The 1912 Wallis Bear was the early start of the Wallis tractor line, which Massey-Harris acquired in 1928 when it bought the J. I. Case Plow Works of Racine, Wisconsin. A four-cylinder inline engine, radiator cooling, and 30/50 rating made the Bear a respected tractor for its era. The Bear measured nearly 20 feet in length and weighed a hefty 10.5 tons. E. F. "Gene" Schmidt of Bluffton, Ohio, had this giant in his extensive collection of Wallis and Massey-Harris tractors.

1910s Wallis advertisement

The Massey-Harris Challenger, a two- to three-plow tricycle row-crop tractor, debuted in 1936. It was a 26/36 tractor that had adjustable rear wheels for different row widths. It still used the U-channel boilerplate frame of its Wallis ancestors, but was available on rubber or steel wheels. The Challenger featured a streamlined cast grille.

Another transitional tractor between the Wallis and M-H designs was the Twin-Power Pacemaker of 1937. It was a high-compression tractor designed to burn high-octane (68 to 70) gasoline. Its twin-power feature produced two power output ratings by

Right and inset: 1914 Wallis Cub

Wallis reduced its Bear model tractor size by half for the new 1914 Cub. Still of more than adequate dimensions, the Cub's four-cylinder 840-cubic-inch engine had a 26/44 rating, or enough to pull five to six plows. Weight of the Cub was only 4.2 tons. The patented U-shaped boilerplate frame enclosed the crankshaft, clutch, transmission, and eventually the final drive. Massey-Harris collector-restorer Dick Carroll of Alta Vista, Kansas, found this Cub in Perryton, Texas, in 1993 and restored it the same year.
Inset: The Wallis Cub used this direction finder over its front wheel to remind the driver which way the wheel was pointed.

1914 Wallis Cub

In an effort to reduce wear on its final-drive components, the Cub's exhaust gases were blown over the pinion gear at its contact with the ring gear on the tractor's drive wheel. An occasional shot of blow-by oil probably gave the tractor some lubrication. The next Wallis model enclosed the final drive.

Below and opposite: **1916 Wallis Cub Junior**

Half the size of its "Papa" Cub, the Wallis Cub Junior, or Model J, of 1916 generated 15 drawbar horsepower and 27 belt horsepower with its 4.5x5.75-inch engine. The Model J had a two-speed transmission, Hyatt roller bearings, and a belt pulley tucked behind its left drive wheel. Wallis was able to enclose the final drive within the boilerplate bathtub frame. Starting the Cub Jr. involved cranking it with a ratchet lever on the pulley shaft. Construction of its single-wheel front end left no room for a conventional crank. Dick Carroll of Alta Vista, Kansas, also restored this Wallis Model J for his Massey collection.

1918 Massey-Harris MH-1
The 1918 Parrett, designated the model MH-1, was once sold by Massey-Harris. The tractor originated in Chicago, Illinois, but was assembled by Massey in Weston, Ontario, from 1918 to 1923. Dallas Center, Iowa, Massey-Harris fan Lawrence Myers found and restored this early machine. Massey also sold MH-2 and MH-3 versions.

governing the engine at either 1,200 or 1,400 rpm. The row-crop tricycle Challenger of 1938 also had the Twin-Power feature. It was the last of the Wallis-type tractors.

MASSEY BUILDS ITS OWN TRACTORS

In 1938, Massey brought out the red M-H 101 row-crop with a new cast frame and sheet metal streamlining. It was powered with a Chrysler-made six-cylinder L-head engine, designed to thrive on high-test gasoline. It was also available in a standard, or wheatland, version. Both machines had the Twin-Power feature.

A wartime model, the Massey-Harris 81, was pressed into service on Royal Canadian Air bases as an aircraft tow tractor. The row-crop Model 81 was built from 1941 to 1948 and a standard version from 1941 to 1946. The M-H 81 was powered by a Continental four-cylinder 3x4.375-inch engine. Only about 6,000 were made. It was followed in 1947 by the Massey-Harris

Model 20, a one- to two-plow tractor designed for the lighter work on the farm.

Another 1947 entry was the 30, a two-plow tractor available in row-crop or standard-tread and equipped with a five-speed transmission. Its engine was a Continental L head turning at either 1,500 rpm for drawbar work or 1,800 rpm for belt applications. The 30 was upgraded to the Model 33 and the Model 333 during its production from 1947 to 1953; more than 32,000 units were made during that time period.

The Model 30's larger stablemates for the 1947 model year included the three-plow M-H 44, available with a four- or six-cylinder gas engine, or a four-cylinder diesel, and the M-H 55, a four-plow tractor for the big jobs. The 55 had a 60-horsepower engine and was available in gas or diesel. During its production run from 1947 to 1956, about 21,000 Model 55s were built. More than 95,000 Model 44s were sold during its production life from 1947 to 1955.

1927 Wallis 20/30

The 1927 Wallis 20/30 Certified came to its new owner with papers signed by company officers guaranteeing that it would perform as claimed. The four-cylinder 4.375x5.75-inch engine ran at 1,050 rpm to reach the 20/30 rating. Two front wheels of standard tread with automotive-type steering, deep fenders, and a spring seat made it a modern tractor. Wallis 20/30 operators had a seat protected from flying dirt by deep fenders. Its steering wheel was mounted to the far right for a good view of the plow furrow. The tractor could burn gasoline or kerosene and used water injection to control pre-ignition knocking in the engine. Massey sold some Wallis tractors in 1927 before buying the whole line the next year. Richard Groves of Webster City, Iowa, restored this historic tractor.

1930 Wallis 12/20

This 1930 12/20 model was among the Wallis tractors sold by Massey-Harris after it bought the J. I. Case Plow Works Company in 1928. This smaller Wallis model was powered by a 3.875x5.25-inch four-cylinder valve-in-head engine. It had a three-speed transmission and weighed about 3,400 pounds. It evolved into the standard-tread M-H Pacemaker of 1936. Carlene and Richard Meyer of Dudley, Massachusetts, restored this Wallis.

1930 Massey-Harris GP
Designed and built by Massey-Harris as its first row-crop tractor, the 1930 Model GP came closer to being a four-wheel-drive tillage tractor. The 15/22 general-purpose tractor was available in four different wheel widths for different row spacings. Drop gearboxes at each of its four wheels kept the main axles high for row-crop clearance. The rear axle oscillated over uneven ground at a pivot point at the rear differential. The two front wheels steered through sealed universal joints. Raymond Krukewitt of Sidney, Illinois, invested a winter of spare time restoring this machine.

The small Pony tractor was introduced by M-H in 1947. Smallest in the Massey line, the Pony could pull one small plow or do many other light duties on the farm. Its 2.375x3.5-inch Continental engine spun at 1,800 rpm to deliver about 10 horsepower on the drawbar and 11 at the PTO. The Pony was built between 1947 and 1954 in Canada, with more than 27,000 made.

FERGUSON JOINS THE FOLD

The year 1953 marked the start of many changes at Massey as the merger with Harry Ferguson created Massey-Harris-Ferguson. Some of the Massey-Harris tractors were updated and continued in production. Initially, the tractors were sold through either Massey-Harris or Ferguson dealerships. Later, the dealerships were consolidated.

1938 Massey-Harris Challenger
Designed as a tricycle-type tractor for row-crop applications, the 1936 Challenger got that job done. Massey-Harris specialist Kurt Kelsey of Iowa Falls, Iowa, restored his beautiful 1938 red Challenger. Its rounded cast-iron grille with attached cutout hood lends an early streamlining touch to the tractor. This 26/36 tractor was the last of the Wallis-type machines with their U-shaped boilerplate frames.

The M-H 33 became the 33 RT gasoline. Its Nebraska tests showed 35.54 horsepower on the drawbar and 39.52 belt horsepower. It was also available as a diesel. The 33 RT was produced between 1953 and 1957. The M-H 44 Special was a gasoline standard tractor that delivered 43.58 drawbar horsepower and 48.95 belt horsepower. It too was available with a diesel engine. The Massey-Harris No. 16 Pacer of 1954–1957 was a larger version of the Pony with 17 drawbar horsepower.

Massey's upgraded Model 555 gasoline and diesel tractors replaced the Model 55 in 1955. The 555 offered Depth-O-Matic, the Massey-Ferguson equivalent of the Ferguson system of hydraulic lift and implement control. On the Ferguson-derived machines, the hydraulic system was called Hydramic Power.

Under the new merger, M-H continued to build Ferguson tractors as well. The Ferguson Model TO-35 of 1955 was built in Ferguson's Detroit plant. The TO-35's 2,000-rpm 134-cubic-inch

1938 Massey-Harris 101
Pre-war chrome trim graces the new rounded lines of the 1938 Massey-Harris Model 101. This new M-H streamlined design tractor was powered with a high-compression six-cylinder Chrysler L-head gas engine. Richard Prince of Conover, Ohio, pulled this tractor from a collapsed barn and then fixed it up. Louvered engine covers helped vent the engine compartment.

1941 Massey-Harris buyer's guide

1939 Massey-Harris 101

Electric lights and a starter are appreciated options on this 1939 standard-tread version of the Massey-Harris 101. Deep fenders cover the big drive wheels. The M-H Twin Power feature provided two governed engine speeds for different kinds of power application. Drawbar work was accomplished at 1,500 rpm, and belt pulley applications permitted a power setting of 1,800 rpm. Dick Carroll of Alta Vista, Kansas, collected and restored this 101.

1942 Massey-Harris 81
Production limits imposed by World War II kept down the production numbers of this 1942 Massey-Harris Model 81. The light two-plow machine was equipped with a wide front-adjustable axle for row-crop work. Its power came from a 124-cubic-inch 3x4.375-inch bore-and-stroke Continental four-cylinder engine. Ronald L. Hoffmeister of Altamont, Illinois, found this tractor near his home and rebuilt it into this shining collectible.

Continental engine put out 30 horsepower on the drawbar with its six-speed transmission, making it a three-plow machine. The Ferguson TO-35 later became the MF-35, painted Massey-Ferguson red and gray and one of the best sellers in the new M-F line.

The year 1956 brought out the Ferguson gas 40 and the Massey-Harris gas 50. Identical under the skin, these tractors had the same 134-cubic-inch Continental engine and put out just over 30 drawbar horsepower. LPG versions and diesel versions of the

M-H 50 were offered, and the model was made from 1956 to 1965. In 1956, Massey also updated the 33 to the 333 and offered it in gasoline, diesel, or LPG versions. The M-H 444 of 1956 to 1958 became available with LPG and diesel engine options. Both the 333 and 444 offered Depth-O-Matic.

By 1958, Massey-Ferguson was offering tractor models ranging from 25 up to 60 horsepower with diesel or gasoline engines. Larger tractors followed.

1949 Massey-Harris Pony
Smallest of the Massey-Harris tractors was the Pony. Aimed at truck farms and small-acreage vegetable growers, the Pony delivered about 10 horsepower to the drawbar from a four-cylinder 1.375x3.5-inch 1,800-rpm engine. Cast-iron wheel weights kept the little Pony pulling. Introduced in 1947, the Pony was made until 1954 in Woodstock, Ontario. A slightly larger model with 15 horsepower was the Pacer. Don Rogers of Atlanta, Illinois, restored this pretty 1949 Pony.

1951 Massey-Harris 44 Diesel
Diesel power was an option in the Model 44 Massey-Harris machines. The Model 44 was a popular tractor in post-war North America, when diesel engines were seeing early farm service. Professional tractor-restorer Dan Peterman of Webster City, Iowa, restored this standard-tread tractor for his own Massey collection. It is a 1951 model with a four-cylinder 260-cubic-inch diesel.

FOR MORE POWER, SPEED, COMFORT AND ECONOMY

Make it a
MASSEY-HARRIS

Two-Plow Model "20"

Full Two-Plow Model "30"

Four-Five Plow Model "55"

Three-Plow Model "44-6"

Three-Plow Model "44"

THERE'S A MASSEY-HARRIS TRACTOR FOR EVERY SIZE FARM!

1950 Massey-Harris 44 Vineyard
The narrow design of the Model 44 Vineyard kept it out of the grape vines.

1950s Massey-Harris advertisement

PERKINS DIESEL ADDITION

Diesel engine manufacturer F. Perkins Ltd., of Peterborough, Ontario, was bought by Massey-Ferguson in 1959. Perkins had been supplying diesel engines to Massey-Harris since 1947 when its engines first powered the M-H 44.

A three-cylinder Perkins diesel was offered in the MF-35 of 1959–1965. Displacement on the three-cylinder diesel engine was 152.7 cubic inches, and it provided 33.02 drawbar horsepower and 37.04 belt horsepower in Nebraska tests. Bore and stroke were 3.6x5 inches, and its operating speed was 2,000 rpm.

A 60-horsepower MF-85 LPG was announced in 1959. Its 242-cubic-inch four-cylinder 3.875x5.125-inch Continental engine developed 56.47 drawbar horsepower and 62.21 belt horsepower via an eight-speed transmission. A gasoline version had similar tests results. The MF-85 was made from 1959 to 1962.

In 1960, the MF-88 diesel debuted. It was nearly the same as the MF-85, but was powered by a 276.5-cubic-inch Continental four with a 4x5.5-inch bore and stroke. Its tests showed 55.54 horsepower on the drawbar and 63.31 belt horsepower. Also in 1960, the MF-65 diesel was introduced with a 203.5-cubic-inch four-cylinder Perkins diesel engine. It produced 42.96 drawbar horsepower and 48.59 belt horsepower in Nebraska tests with its 3.6x5-inch bore-and-stroke engine running at 2,000 rpm. It had a six-speed transmission.

Massey-Ferguson bought Italian tractor maker Landini in 1960. Landini made only diesel-engine tractors from its beginning in 1910. M-F sold some of the blue-painted Landini models in the United States and later manufactured some M-F models in the Landini plants.

1954 Massey-Harris 44 High-Crop
Massey-Harris addressed the needs of specialty crop growers with this high-stepping 1954 Model 44 High-Crop. It was powered by a 277-cubic-inch Continental engine that produced 43.58 drawbar horsepower when tested at Nebraska. Larry Maasdam of Clarion, Iowa, owns this tractor that was once available for work in staked vegetables, sugar cane, or other tall crops.

1953 Massey-Harris Colt
The Massey-Harris Colt replaced the Model 22 in the red-machine stable in 1952. This 1953 Colt has the new Depth-O-Matic three-point hitch for attaching rear-mounted implements. Special curved front-end weights help counterbalance the rear-mounted implements. The two-plow Colt harnessed energy from a 124-cubic-inch four-cylinder gas engine making power at the 20-horsepower level. A slightly bigger "horse," the Massey-Harris Mustang 1953 forced the Colt "out to pasture." Glen Mlnarik of Howells, Nebraska, restored this Colt.

LOOKING TO THE FUTURE

Meeting its competition with both larger and smaller tractors kept Massey-Ferguson on its toes for the next twenty years. In the late 1970s, Massey losses began to mount as interest rates soared and farm equipment sales lagged. Massey began selling off divisions and consolidating production by closing plants. The early 1980s were financially brutal, and business conditions worsened.

In 1993, Massey-Ferguson's parent company, Varity Corporation, sold North American Massey-Ferguson distribution

rights to AGCO (Allis-Gleaner Company). AGCO had been formed in 1990 to buy the Deutz-Allis Corporation from its German owner, Klockner-Humboldt-Deutz. In 1994, Varity Corporation sold its Massey-Ferguson division to AGCO, but kept the Perkins engine operation. Today, Perkins engines are owned by Caterpillar, Inc., and Massey-Ferguson continues operations under the AGCO umbrella, making and selling a complete line of farm tractors and combines.

1954 Massey-Harris Mustang
The Mustang model of 1953 was a bit more powerful than the earlier Colt. Some of its features included a velvet ride seat and a three-point hitch for easy hydraulic control of mounted implements. This 1954 Mustang was built soon after Massey-Harris bought Ferguson and the company became Massey-Harris-Ferguson in August 1953. The name was soon shortened to Massey-Ferguson, Inc. As a result of the merger, the Massey-Harris two-plow Colt and Mustang models were dropped and replaced with two-plow Ferguson Model TO-20 tractors. This Mustang is owned by Paul Lehman of Perry, Iowa.

1950s Ferguson Model 30 advertisement

Below: 1955 Massey-Harris 55 LPG
A big standard-tread tractor for the western wheat country, the Massey-Harris Model 55 was a four- to five-plow workhorse. The Model 55 featured a 382-cubic-inch four-cylinder engine that put 63 horsepower on the drawbar. This 1955 Model 55 Massey, equipped to burn LPG, somehow found its way east to Pennsylvania and into the M-H tractor collection of John Dudkewitz of Cochranville, Pennsylvania.

Next page: 1956 Massey-Harris-Ferguson 333
This was the first of the Massey-Harris-Ferguson tractors: the Model 333. It was basically a Massey-Harris model with a new MF paint scheme, which included a gold color on the engine. This 1956 row-crop three-plow tractor is fitted with the adjustable wide front for row-crop use. The 333 evolved from the Model 30 two-plow series first offered by Massey-Harris in 1947. Professional tractor restorer Paul Lehman and his wife Rose of Perry, Iowa, put the sparkle back in this half-century-old machine.

Chapter Nine
Minneapolis-Moline

1938 Minneapolis-Moline UDLX

Way ahead of its time, the 1938 Minneapolis-Moline Model UDLX Comfortractor led the tractor industry by nearly thirty years in providing enclosed-cab operator comfort. The all-weather cab with heater, cigar lighter, roll-down rear windows, venting front widows, windshield wipers, passenger seat, radio, spotlight, electric lights, starter, horn, and full fenders nearly doubled the price of the basic tractor. This groundbreaking UDLX was restored by Don Kingen of McCordsville, Indiana.

A rugged northern prairie flavor was apparent in the tractors produced by the new Minneapolis-Moline Power Implement Company after its organization in 1929. Three firms with long histories in making tractors and agricultural implements consolidated to form the new company: The Minneapolis Steel & Machinery Company and the Minneapolis Threshing Machine Company, both of Minneapolis, Minnesota, and the Moline Implement Company of Moline, Illinois. The machines they eventually built together were powerful and strong, hung on at heavy loads, and were ready for the big prairies of the northwest.

Minneapolis Steel brought its Twin City tractor-building experience to the merger, while the Minneapolis Threshing Company added its Minneapolis tractors, combine, threshers, and corn shellers into the fold. Moline Implement, formerly Moline Plow Company, contributed its long line of plows, discs harrows, and other farm implements.

THE FLYING DUTCHMAN

Moline Plow's predecessor company, Candee & Swan, was an early plow-making competitor of John Deere. The two companies struggled in court in 1867 over the right of each firm to call their product the Moline Plow. Deere won in district court, but the ruling was overturned by the Illinois Supreme Court in 1871, and Candee & Swan soon named itself the Moline Plow Company. After buying a sulky plow design in 1883 known as the Flying Dutchman, Moline Plow adopted the name and a Flying

Dutchman emblem as its trademark. It discontinued use of the Flying Dutchman trademark at the onset of World War I and never reinstated it.

By 1919, Moline Implement had developed and was marketing an advanced articulated general-purpose tractor called the Moline Universal. The end for the Universal came in 1923 during the tractor price wars instigated by Henry Ford and joined by International Harvester. All farm equipment makers were experiencing hard times in a post–World War I recession, and many of them stopped or paused production of some of their products.

THE TWIN CITY TRACTOR

Minnesota Steel and Machinery Company was organized in 1902 to build steel structures, but by 1903 the company was making stationary steam engines, and in 1904 it started producing early gasoline engines. MS&M subcontracted the manufacture of equipment for other firms, including J. I. Case Threshing Machine Company. In 1912, MS&M agreed to build 500 early Case 30/60 gas tractors. The company also contracted in 1913 to build 4,600 Bull tractors, with Bull furnishing the engines for the briefly popular three-wheel machine.

1917 Minneapolis 35/70
Built in the tradition of heavy steam engines of its day, this towering 1917 Minneapolis 35/70 gas tractor still shakes the ground with its 11.25-ton mass. Its big four-cylinder cross-mounted engine has a bore and stroke of 7.25x9 inches and produces 35 drawbar horsepower and 70 belt horsepower. Minneapolis Threshing Machine Company of Hopkins, Minnesota, made the tractor to run big threshing machines and plow the wide-open northern prairies. It takes four big steps to climb into the 35/70 cab. KLS Implement of Elrosa, Minnesota, owns and cares for the tractor. KLS includes Steve Korf, Gene Loxtercamp, and Tom Siefermann.

1919 Minneapolis-Moline Universal D

With built-in electric lights and starter, the 1919 Moline Universal Model D was very modern for its day. Its owner could buy implements from the Moline Plow Company or modify his own horse-drawn equipment. The tractor was rated at 9/18, but demonstrated 17.40 drawbar horsepower and 27.45 belt horsepower in Nebraska tests from its four-cylinder 3.5x5-inch 1,800-rpm engine. Long operating rods connected the Universal driver with his machine. The Universal Model D was made from 1918 to 1923. Concrete wheel weights helped traction and lowered the D's center of gravity. Jim Jonas of Wahoo, Nebraska, had help from his father, brother, and uncle in restoring this machine.

Minneapolis Steel had also developed its own line of Twin City tractors in the meantime, and in 1915 it offered four different power sizes as the TC-15, TC-25, TC-40, and the monster TC-60. The company's 1913 Model 60/110 (later the 60/90) had a six-cylinder engine with a 7.25x9-inch bore and stroke. The 60/90 was a big tractor, weighing nearly 14 tons. By 1924, the big tractors were dropped in favor of lighter tractors as a result of the precedent-setting Fordson from Detroit, Michigan.

The Twin City Model 12/20, introduced in 1919, was a standard-configured tractor with a unit-frame construction. Four valves per cylinder—two intake and two exhaust with twin cams—was an innovation that improved the engine's breathing. More powerful models followed, with a 20/35 tractor in late 1919; an improved 17/28 in 1924; a stronger tractor, the 27/44 in 1924; and the 21/32 model in 1929.

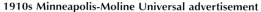

1910s Minneapolis-Moline Universal advertisement

Left:
1924 Twin City 12/20 Wheel
Long cast wheel lugs gave the Twin City 12/20 extra traction for working soft delta soils.

Opposite:
1924 Twin City 12/20
Also surprisingly modern for its time was the 1924 Twin City 12/20, made by Minneapolis Steel & Machinery Company. Its sixteen-valve four-cylinder engine was the first such engine known on a farm tractor. Each cylinder had two intake and two exhaust valves. Leslie B. Moffatt of Brighton, Tennessee, restored this Twin City 12/20 after he had released it from the clutches of a large persimmon tree.

THE MINNEAPOLIS TRACTOR

Minneapolis Threshing Machine Company was a veteran producer of steam traction engines and threshing machines. The company had a history that dated back to 1874, when it was founded as the Fond du Lac Threshing Machine Company of Fond du Lac, Wisconsin. In 1910, Minneapolis Threshing began selling the Universal 20/40 two-cylinder opposed gas tractor, made by the Universal Tractor Company in Stillwater, Minnesota.

MTM went on to develop its own line of gas tractors, including its 1911 Model 25/50 with a four-cylinder engine. Other MTM tractors were a 40/80 (later rated as a 35/70) in 1912, a 20/40 in 1914, and a 15/30 in 1915 (later designated a 12/25 after Nebraska tests). Smaller models followed in the 1920s, including MTM's first tractor with a unit frame, the 1922 crossmounted four-cylinder 17/30. Its radiator tank casting proudly identified it as "The Minneapolis Tractor." Another interesting feature was a belt idler

1930 Twin City KT

The 1930 Twin City Model KT (for Kombination Tractor) was a three-row row-crop specialist with this mounted cultivator. The 14/23 KT became the first tractor to be sold by the new Minneapolis-Moline Power Implement Company after its creation in 1929. The firm combined the following companies: Minneapolis Steel & Machinery, Minneapolis Threshing Machine Company, and Moline Plow Company. Computer software engineer Curtis Rink of Wichita, Kansas, restored this tractor. He collects M-M tractors and farm equipment.

pulley attached to the axle of its right front wheel. Two versions of the model were designated the 17/30A and 17/30B.

The B version, with its 4.875x7-inch bore and stroke, was a trendsetter. The long-stroke concept from Minneapolis Threshing, giving an engine superior lugging power at low revolutions, was passed on to the new Minneapolis-Moline designs of the 1930s. That long-stroke concept was kept alive by M-M as long as it made tractors and engines. It gave M-M engines the ability to hang on and recover to governed speed on heavy loads, like while performing heavy belt work on threshers and corn shellers.

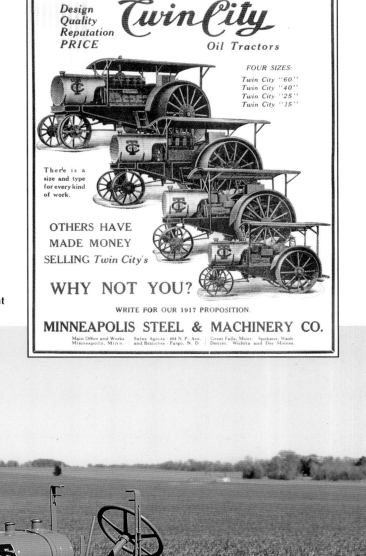

1917 Twin City Model 60, 40, 25, and 15 advertisement

1930 Twin City KT
One crop row was straddled by the KT as it cultivated three rows. It later became a two-row machine.

The Minneapolis-Moline Universal J of 1935 was the tricycle row-crop version of a tractor also offered as a Standard J and as a larger Universal M. Still painted gray with red wheels, the J models came with a 3.625x4.75-inch four-cylinder engine. The M engine was a 4.25x5-inch four-cylinder design. M-M collector Dale Nafe of Pierson, Iowa, restored this Universal J.

MINNIE MO ON THE GO

The new Minneapolis-Moline Power Implement Company had a challenging task ahead in 1929. It needed to sell off old inventory from its three merged companies and quickly develop new products for a rapidly changing market—all in the face of a deepening recession. M-M was up for the game.

First out in 1930 was the KT Combination Tractor general-purpose machine.

Designed with a bowed front axle and extra under-axle crop clearance, the two-plow 14/23 tractor was a quadruple-duty machine built to cultivate two to three rows, pull tillage equipment, power belt applications, or run PTO-powered equipment. It stayed in the M-M line through 1934 and then was upgraded to the Model KTA, which was made until 1938. Orchard versions of the KT were also produced. The Twin City and M-M names were both cast in the radiator tops of their tractors from 1931 to about 1938; after that, only the M-M trademark appeared.

The first M-M tricycle row-crop was the Universal MT 13/25, introduced in 1931. The 4,860-pound machine was configured a lot like the IHC Farmall, with drop-box drive-wheel gearing to

1935 Minneapolis-Moline Universal J
The J air cleaner, with its high intake stack, rode up front for cleaner air.

1937 Minneapolis-Moline ZTI Industrial
Once equipped with a front loader, this 1937 Minneapolis-Moline Model ZTI Industrial sports the Prairie Gold and red trim colors of the new Visionlined M-M line. Referred to as "red noses" by collectors because of their red radiator grilles, the new Prairie Gold line was introduced with this streamlining. Electric lights and starter were popular options. Big rear wheel weights kept the rear wheels on the ground when the loader bucket was full. KLS Implements of Elrosa, Minnesota, owns this ZTI.

1937 Minneapolis-Moline ZTI Industrial
The Z model had a four-cylinder 3.625x4.5-inch engine that put out 20.98 drawbar horsepower.

provide under-axle crop clearance. Front wheel steering was controlled by a side rod turning the front wheels by a projecting lever, similar to the Case row-crop tractors. The 13/25's four-cylinder 4.35x5-inch bore-and-stroke engine was designed to run at 1,150 rpm. It was updated to the Model MTA in 1935 with an optional high-compression engine, designed to burn 70 octane gasoline. Another option was a 10-mile-per-hour high-speed gear, which was popular in tractors equipped with rubber tires.

GOODBYE ROOSTER ROOST

The new M-M Universal J row-crop tractor introduced in 1935 pioneered an F-head engine with exhaust valves in the block and the intake valves in the head. M-M claimed the valves would be better cooled with the additional water jacketing surrounding them. The 3.625x4.75-inch engine gave the Model J power to pull two plows. A five-speed transmission was in line with the times. In fifth gear, it could scoot the tractor down the road at 12.2 miles per

1937 Minneapolis-Moline ZTU

The compression ratio on the Minneapolis-Moline ZTU's kerosene engine could be raised to enable it to burn gasoline. M-M ads claimed the Z's engine could be overhauled while sitting on a milk stool beside the tractor. From smallest to largest, the Visionlined M-M tractor models included the R, the Z, and the U, as well as the big GT standard-tread tractor. Virden Smith and his son Biron of Findlay, Ohio, made a family project out of restoring this 1937 Minneapolis-Moline Model ZTU.

Above and right: **Minneapolis-Moline Visionlined Series brochures**

hour if equipped with rubber tires. High-octane versions of the J were available in 1936 as were J orchard models.

The Model J resembled the earlier Model M, but its steering was via a worm-and-sector gear in its front pedestal, turned by a steering rod that was mounted in the tractor frame. The rod turned on ball bearings and was connected to the angled steering column with a universal joint; its rooster roost was gone.

Like other second-generation row-crop tractors, the J adopted larger diameter rear wheels attached to long, single-piece rear axles driven directly from the differential. Hub clamps on the wheels allowed infinite wheel width adjustment. The J's air cleaner, mounted high in front on the steering pedestal, was a distinctive feature. It was placed to pull in dust-free air high above

the tractor's front wheels. The Universal J was also made in a standard-tread version.

GOING GOLD

Bright colors were beginning to adorn farm tractors as their makers pushed past the Depression and aimed at brighter times. The drab grays of the M-M tractors became history in 1937, when M-M's new Visionlined Model Z was introduced. An appropriately bright Prairie Gold color adorned the new streamlined machines. Neither yellow nor orange, the color lay somewhere in between and seemed to reflect the color of ripe wheat in the late summer sunshine. Picking up the red from the wheels of the previous tractor models, M-M also used the red as a trim color on the radiator

1938 Minneapolis-Moline UDLX
Because UDLX's revolutionary features put the tractor's price out of reach for many farmers, fewer than 150 sold, making the tractor a red-hot collector's item today. Paul and Kay Weiss of Reinbeck, Iowa put this one back together. Under its slick exterior is the rugged three- to four-plow M-M Model U standard tractor with 30.86 drawbar horsepower. On the road the UDLX can roll at 40 miles per hour, if courage and the road surface permit. Rural mail carriers in northern states loved to use this tractor for winter mail delivery.

1938 Minneapolis-Moline UTU
The Minneapolis-Moline Model UTU was the big row-crop tractor in the M-M line. It stayed in production until 1953. The UTU's four-cylinder 4.25x5-inch engine showed 30.86 drawbar horsepower and 38.12 belt horsepower when tested at Nebraska in 1938. Longtime M-M collector Roger Mohr of Vail, Iowa, restored this beautiful 1938 Model UTU. This model stayed in production until 1953.

grille to mark the new line of tractors now affectionately called Red Noses by classic tractor buffs.

The tapered Visionlined hood of the Z gave its row-crop operator better visibility for cultivating. Also new on the Z was an engine design that permitted it to run on 70-octane gasoline or the cheaper fuels, such as kerosene and distillates. A detachable head with adjustable protrusions in the combustion chamber permitted the compression ratio to be changed for use of either gasoline or the low-cost fuels. The Zs were soon available as standard-tread models as well. The ZTU was a two- to three-plow tractor that produced 20.98 drawbar horsepower and 27.95 belt horsepower in its 1940 Nebraska tests. Its four-cylinder engine had 3.625x4.5-inch bore and stroke and ran at 1,500 rpm.

The famous Comfortractor 1938 M-M model introductions were big news. The beautiful new Prairie Gold three- to four-plow Model U enclosed in its futuristic all-weather cab with sweeping fenders was a crowd pleaser. The Model UDLX (for deluxe) had just about everything an operator might want. Its all-weather cab had roll-down side windows, full venting front windows with wipers, a radio, a top-mounted spotlight, a jump seat for a passenger, a heater, a foot accelerator, and even a horn. Electric lights, a starter, and a five-speed transmission gave the machine added performance, too—up to 40 miles per hour if the road surface was smooth. The UDLX became known as the Comfortractor.

But for all its gloss and appeal, the UDLX was not fully appreciated for nearly fifty years, until it became a favorite of

1941 Minneapolis-Moline R
The 1941 Minneapolis-Moline Model R row-crop tractor with enclosed cab was another attempt by M-M to aid operator comfort. But for a tractor that was used for crop cultivation during many weeks in the hot summer, it lacked what it most needed: air-conditioning. Like the big UDLX, the R sold too poorly to be a success. An operator could watch the tractor's front-mounted two-row cultivator through the lower front cab windows. Paul and Kay Weiss of Reinbeck, Iowa, bought this R to accompany their UDLX Comfortractor.

classic tractor collectors. Only about 150 UDLXs were made between 1939 and 1941, with perhaps only half of them surviving, UDLXs today can sell for many, many times their original price of $1,900.

Ironically, cost was the major stumbling block to the UDLX's original sales success. The added comfort equipment nearly doubled the price of the same tractor without the frills. Under its fancy skin, the UDLX was a hefty three-plow tractor that was used by some farm owners and custom combine operators who appreci-

ated its speed and comfort in long road pulls between jobs. Some rural mail carriers in the northern prairie states also praised the UDLX's snug warm cab and go-anywhere capabilities when the north wind howled and the snow drifts were deep.

The Model U engines were four-cylinder 4.25x5-inch units that delivered 30.86 drawbar horsepower and 38.12 belt horsepower at 1,275 rpm in 1938 Nebraska tests. The U, without the comfort features, was available as a row-crop tricycle and standard-tread version.

1941 Minneapolis-Moline R
The double rear doors aided entry into the Model R cab. About 400 of the cab-equipped Rs sold.

Another cab-enclosed tractor, M-M's two-plow tricycle-front row-crop R, was available from 1939 to 1941. It was less expensive than the UDLX, but it did not sell well. Only about 400 of the Comfort-Cab-equipped Rs were sold. The RTU, tested at Nebraska in 1940, delivered 15.58 drawbar horsepower and 20.49 belt horsepower from its 3.625x4.5-inch engine turning at 1,400 rpm. The cab-equipped R had a front-mounted two-row cultivator. The belt pulley could be used, but its clearance from the cab was minimal. In 1941, the two-plow R was available either in a tricycle or standard wheel-tread configuration. It would be nearly thirty years before fully enclosed tractor cabs became popular on U.S. farm tractors.

Also new in 1939 was the M-M Model GT, a standard-tread or wheatland tractor with four- to five-plow power. The GT was tested at Nebraska in 1939 where it produced 36.27 drawbar horsepower and 48.93 belt horsepower; these ratings came from a four-cylinder 4.625x6-inch engine turning at 1,075 revolutions.

Next page: **1941 Minneapolis-Moline UTS LPG**
Minneapolis-Moline pioneered the use of LPG, or liquefied petroleum gas, in its tractors in the early 1940s. This 1941 Model UTS was one of the first tractors launched that ran on LPG. This standard-tread model is also equipped with a special LPG vaporizer used to mix the gas with engine intake air. The compressed fuel is stored in the LPG pressure tank that replaces the gas tank on this tractor. LP fuels burned clean, lengthened engine life, and helped maintain full power. Billy Montgomery of Columbia, Missouri, restored this big M-M.

1949 Minneapolis-Moline ZAU
New styling and new features were added to this post-war row-crop Minneapolis-Moline Model ZAU, including a five-speed transmission and a new yellow grille and hood design. James Garrod of Sheridan, Iowa, restored this 38-horsepower tractor with many hours of work late at night and on weekends. It was delivered new with electric lights and a starter.

1951 Minneapolis-Moline R
Still looking much like the original Visionlined Model R of 1937, this 1951 Minneapolis-Moline Model R had new disc brakes, a four-speed transmission, and a 165-cubic-inch four-cylinder gas engine that supplied 23.90 drawbar horsepower and 27.09 belt horsepower. The two-row, two-plow tractor was restored to this sparkling new condition by Tom Humbert of McConnellsburg, Pennsylvania.

PIONEERING LPG POWER

Liquified Petroleum Gas, a pressurized mixture of volatile petroleum waste gases, including butane and propane, was bottled for fuel uses in the mid-1930s. Using LPG as engine fuel was pioneered by Minneapolis-Moline engineers, as M-M was the first tractor manufacturer to offer factory-built LPG tractors in 1941. Other manufacturers soon followed, and LPG became an alternative to gasoline, kerosene, or distillates. Down the road, diesel-fueled tractors replaced most LPG burners.

World War II interrupted model introductions at M-M, as it did at other tractor manufacturers. Some of M-M's defense contracts were for utility towing vehicles based on some of its tractor models. One of these, a four-wheel-drive unit based on the U series, is said to have inspired the name "Jeep," later used as the name for the lightweight, all-purpose four-wheel-drive military utility vehicle.

Restyling and upgrading the R, Z, U, and G Series began about 1948. The 1948 ZA boasted a flat front with louvered grille painted Prairie Gold instead of the previous red. It came in many different configurations. The general-purpose row-crop version was the ZAU, with twin front tricycle wheels, and the ZAN was a single-front-wheel version. The ZAE had an adjustable wide front and rear tread width, while the ZAS was the standard-tread tractor.

The big Model G was converted to burn LPG in 1950 and designated the GTC. The GTB gas version was produced from 1951 to 1953.

1955 Minneapolis-Moline GB Diesel
Minneapolis-Moline added diesel engines to its tractor line with this 1955 Model GB. Its 425.5-cubic-inch six-cylinder diesel cranked out 55.44 drawbar horsepower and 62.78 belt horsepower at 1,300 rpm when tested at Nebraska. The GB model was also offered in gasoline- and LPG-fueled versions. This big diesel spent its early days working in the wheat fields of western Kansas. Owner Bill Lile of Indianda, Iowa, put this 7,400-pound machine back in shape.

In 1949, the U was the first factory-designed LPG tractor to be tested at Nebraska. The U stayed in production from 1947 to 1959. The UTC, a high-clearance sugar cane version, was available from 1948 to 1954.

The R Series was available with disc brakes and received the same styling change as the Z in the early 1950s. It, too, was available in different wheel configurations.

THE UNI-TRACTOR AND DIESEL POWER

A unique universal power unit for propelling various harvesting equipment was developed by Minneapolis-Moline in 1951. Called the Uni-Tractor Model L, it was designed to mount and power M-M's Uni-Forager, Uni-Harvestor, Uni-Picker/sheller, Uni-Baler, and other powered implements. Its nickname was the Motorcycle because its left front drive wheel, powertrain, and rear wheel lined up in a row with the operator seated top front. The right drive

wheel was at the end of a long, reinforced axle housing with attachment points where the accessory was mounted. A 45-belt horsepower, 206-cubic-inch V-4 engine powered the Uni-Tractor. The V-4 engine was replaced in 1960 with an inline engine of the same displacement. New Idea of Coldwater, Ohio, bought the Uni-Tractor and its equipment line from Minneapolis-Moline in 1962.

Diesel power joined the M-M lineup in the mid-1950s. The first model to sport a diesel engine was the U, but not many U diesels were made. Largest of the M-M diesels was the six-cylinder GB diesel. It turned out 55.44 drawbar horsepower and 62.78 belt horsepower at its Nebraska tests in 1955. The engine displaced 425.5 cubic inches from its 4.25x5-inch bore and stroke. The GB was produced between 1955 and 1959 and also was available in gas and LPG versions.

As the market kept asking for more horsepower, M-M responded with a new tractor line to replace the R, Z, U, and G models. The Powerline Series, launched in 1956, featured bold

1955 Minneapolis-Moline BG
Minneapolis-Moline added to its small tractor line in 1951 when it bought the B. F. Avery Company of Louisville, Kentucky, and its line of small tractors and implements. The B. F. Avery one- to two-row tractor was based on the Cletrac General Model GG, which B. F. Avery bought from Oliver in 1941. The M-M versions of the tractor included models with a single front wheel, wide front-adjustable axle, and tricycle front wheels. Dan Tastad of Troy Grove, Illinois, restored this 1955 Minneapolis-Moline Model BG. Its four-cylinder Hercules 133-cubic-inch engine was capable of 24.12 drawbar horsepower.

styling and a new color, Power Yellow. It was darker and more orange than the old Prairie Gold. Missing from the new 445 and a 335 models introduced in 1956 was the familiar hand clutch. The new models also had live PTO, power steering, and a three-point hitch with draft control. Travel speeds were doubled with a new Ampli-torc lever that allowed on-the-go shifting. The 335 was powered with a 165-cubic-inch Minneapolis engine, and the tractor put out 29.84 drawbar horsepower and 33.5 belt

horsepower in its Nebraska tests. The 335 stayed in the M-M line until 1961.

The M-M 5-Star replaced the UB in 1957. Its 283-cubic-inch LPG engine developed 54.96 maximum horsepower in Nebraska tests, and the 336-cubic-inch diesel version put out 54.68 maximum horsepower. The 5-Star model was made from 1957 to 1961.

In 1959, M-M replaced the 455 with its 4-Star model, available with 206-cubic-inch gasoline, LPG, or diesel engines. The 4-Star had improved hydraulics and new styling. Tests of the gasoline-powered 4-Star showed maximum horsepower of 44.57. The LPG version produced a maximum horsepower output of 45.45. The 335 was replaced with the Jet Star, also restyled. The GB replacement was the G-VI. Its six-cylinder 425.5-cubic-inch LPG engine produced 78.44 maximum horsepower, compared with 78.49 for the diesel version. Several more models of the G Series were made from 1959 through 1974, including more powerful models in both two-wheel- and four-wheel-drive versions.

THE WHITE MERGER

In 1960, the M-5 model replaced the 5-Star. It was to be about the last model change at Minneapolis-Moline. Maximum test power from the M-5 diesel version was 58.15 horsepower from the 336-cubic-inch engine running at 1,500 rpm. The M-5 was made between 1960 and 1963.

In 1963, White Motor Corporation bought the Minneapolis-Moline Company after acquiring Oliver Corporation of Chicago in 1960 and Cockshutt Farm Machinery of Brantford, Ontario, in 1962. In 1969, the three companies merged into a new firm created by White called White Farm Equipment, a wholly owned subsidiary of White Motor Corporation. The Minneapolis-Moline name was used on some tractor models until 1974. After that, they became White tractors and were painted silver.

The big Minneapolis-Moline articulated four-wheel-drive tractors were made after White bought the company. The A4T-1400 was powered by a 504-cubic-inch turbocharged M-M diesel that tested 139 belt horsepower in 1969. It was joined in 1970 by the A4T-1600, powered by a 585-cubic-inch M-M diesel. The A4T-1600 turned out 143 PTO horsepower in its Nebraska tests. It was also available with a 504-cubic-inch LPG engine. Some of these tractors were sold as the Oliver 2455 and 2655 models.

1960 Minneapolis-Moline Jet Star
The 1960 Minneapolis-Moline Jet Star was a 44-drawbar horsepower utility-style row-crop tractor in new two-tone M-M colors. Other 1960 M-M models included the larger 4-Star and the M-5 models. M-M collector and restorer Alvin Egbert of New Bremen, Ohio, put this Jet Star back in order. Egbert's Jet Star has a wide adjustable-front axle for use in row crops, an under-tractor exhaust system, and a three-point hitch and hydraulics.

The financial trauma on U.S. farms in 1980 caused serious tremors in the farm equipment business. That year, White sold its farm equipment division to TIC Investment Corporation of Dallas, Texas. Then Allied Equipment Corporation of Chicago, Illinois, bought White Farm Equipment after WFE had declared bankruptcy in 1985. Allied's purchase included the former Oliver plant in Charles City, Iowa, a parts depot in Hopkins, Minnesota, where Minneapolis-Moline had once been located, and all of White's tooling and inventory.

In 1987, Allied Products combined White Farm Equipment and its New Idea firm into one division, White-New Idea Farm Equipment, and moved its headquarters and parts inventory to Coldwater, Ohio. In late 1993, AGCO of Duluth, Georgia, bought the White-New Idea implement line. That's where Minneapolis-Moline's heritage now rests.

1964 Minneapolis-Moline M604 LPG
Mechanical-drive front-wheel assist helped this 1964 Minneapolis-Moline Model M604 LPG pull its way through soft fields with minimum tire slip. The tractor's 336-cubic-inch four-cylinder LPG-fueled engine gave it about 68 drawbar horsepower to apply to four-bottom plows or other heavy tillage tools. The front wheels could be shifted in and out of gear, depending on the traction needed. Spin-out rims aided in adjusting rear tread width. David Zoller of DuQuoin, Illinois, owns this tractor that shows the 1960s trend toward four-wheel-drive.

Chapter Ten
Oliver Corporation

1936 Oliver Row-Crop 70 HC
One of the slickest tractors to come out to date was the 1935 Oliver Row-Crop 70 HC model. It ran on rubber tires, was smoothly streamlined, was propelled with a high-compression six-cylinder engine, and burned high-test gasoline. The model's HC designation stands for its high-compression engine. The Oliver and Hart-Parr names and the winged plow trademark adorn the Model 70's radiator. The 70 was an immediate success. Minnesota Oliver devotee, Jean Olson of Chatfield, collected and restored this 1936 Model 70, which is like the one he drove for many years doing custom farming. It still has its original tires.

The pioneering tractor firm of Hart-Parr Company of Charles City, Iowa, was an essential part of the new Oliver Farm Equipment Corporation of Chicago when it formed in 1929. Joining Hart-Parr in the merger of four historic farm equipment makers were Oliver Chilled Plow of South Bend, Indiana; Nichols & Shepard of Battle Creek, Michigan; and American Seeding Machine of Springfield, Ohio. The new company used the name of one of its constituent companies and the products and ideas of all four to build a successful new full-line farm-implement maker.

Oliver Chilled Plow Company dated back to 1855, when its Scottish-born founder, James Oliver, invented the chilled steel plow. His patented process produced a plow with a hard surface and a tough core.

Nichols & Shepard Company had been building threshing machines since 1850, when John Nichols built his first thresher in the blacksmith shop he had operated since 1848. Nichols & Shepard was famous for its Red River Special threshing machines.

American Seeding Machine Company had been formed in 1903 with the merger of five farm equipment manufacturers. Among its many products was the Superior grain drill, which stayed in the line under Oliver.

At the time of the merger, full-line companies such as IH and Deere ruled the field. Not to combine and compete could quickly bankrupt a firm. It happened with increasing frequency as the U.S. economy slowed to a crawl in the late 1920s. Mergers and acquisitions became frequent as farm equipment firms faced a fiercely competitive future with drained capital resources.

After the 1929 merger, Oliver's initial challenge was to develop a general-purpose or row-crop tractor and update the aging Hart-Parr tractor line. The Hart-Parr tractors had been developed at the turn of the century and improved over nearly thirty years of manufacture. But by the late 1920s, they were still chugging away as mostly two-cylinder designs with little promise of general-purpose uses.

CHARLES AND CHARLES IN CHARLES CITY

Charles W. Hart and Charles H. Parr began their pioneering work on gasoline engines in the late 1800s while still studying mechanical engineering at the University of Wisconsin in Madison. In 1897, they first formed the Hart-Parr Gasoline Engine Company of Madison. In 1900, they moved their operation to Hart's hometown of Charles City, Iowa, where they

found financing to make gasoline traction engines based on their innovative ideas.

Their efforts led them to build the first factory in the United States dedicated to the production of gasoline traction engines. Hart-Parr is also credited with coining the word "tractor" for machines that had been previously called gasoline traction engines.

The firm's first tractor effort, Hart-Parr No. 1, was made in 1901. It was a big two-cylinder 17/30 machine, boasting a bore and stroke of 9x13 inches. Field testing in 1902 showed the No. 1's weaknesses, so in 1903, Hart-Parr No. 2 appeared. It was a more powerful machine with a 22/45 rating. Hart-Parr No. 3, introduced in 1903, was an 18/30 two-cylinder machine with a 10x13-inch bore and stroke. And it worked!

In fact, one No. 3 Hart-Parr tractor worked hard for seventeen years for George H. Mitchell, who lived just south of Charles

1925 Hart-Parr 12/24
Hart-Parr, an early tractor maker based in Iowa, made this 1925 Model 12/24 at about the same time as the last steam engines were being produced by other makers. This tractor was called a "new" Hart-Parr for major changes that made it a lighter machine. Its two-cylinder horizontal engine had a 5.5x6.5-inch bore and stroke and burned kerosene. Lyle and Kyla Spitznogle of Wapello, Iowa, collected and restored this 1925 Hart-Parr.

City. Mitchell bought the tractor in August 1903 for $1,580 and used it until 1919. In 1924, he sold the No. 3 back to Hart-Parr, and it was used in various historical and advertising displays for Hart-Parr and Oliver. In 1960, Oliver gave it to the Smithsonian Institution in Washington, D.C., where it was displayed until 1991 when it was loaned to Lake Farmpark museum at Kirtland, Ohio.

In weight, scale, and performance, the early Hart-Parr tractors resembled their steam engine competitors—not unusual, since the company focused on that market. The big machines' weight

1910s Hart-Parr Little Devil advertisement

1925 Hart-Parr 12/24
Exhaust blows out the Hart-Parr 12/24's small pipe in the front of tractor, just above the axle and below the radiator. This design distanced the operator from the tractor's engine noise and smoke.

Oliver 18/27

The new Oliver Farm Equipment Corporation introduced its first tractor, a tricycle row-crop, in 1930. With engineering help from both Oliver and Hart-Parr, the new Oliver 18/27 row-crop machine was powered with an inline vertical four-cylinder engine. It was equipped with a single, large front wheel. Its rear-drive steel wheels were advertised as "power on tip toe." Richard and Dana Kuper of Charleston, Arkansas, found and restored this antique Oliver to pristine condition.

was measured in tons, and the 40/80 machine built from 1908 to 1914 topped out at 18 tons. The 60/100 machine of 1911 to 1912 came out at a massive 26 tons. Even the smaller horsepower rigs weighed up to 10 tons.

LIGHTWEIGHT LITTLE DEVIL

The Hart-Parr Little Devil of 1914–1916 was among the company's first tries at building a lighter tractor aimed at replacing horses on the farm. It had an unusual design, including one centered rear drive wheel with two wheels in front. Its two-cylinder two-cycle engine was company-rated at 15/22. Ads for the machine called it "the tractor that fits every farm," and emphasized that the Little Devil could be bought for the price of three or four horses. Some 1,000 were made, but the Little Devil apparently didn't replace enough horses and was gone from the Hart-Parr lineup by 1916.

The "New Hart-Parr" of 1918 took better aim at the emerging lightweight tractor market. The 12/25 tractor employed a two-cylinder horizontal engine with a 6.5x7-inch bore and stroke and was of more conventional design. The 15/30 Type A of 1918–1922 was upgraded to a Type C in 1922 and then to a Type E in 1924–1926. A smaller 10/20, designated the Model B, was produced in 1921. The 10/20 Model C of 1922–1924 had a slightly larger engine.

Four-cylinder power in the 22/40 and the 28/50 models was achieved by widening the tractor frames and fastening two smaller, two-cylinder engines together side by side. All were horizontal low-speed engines. Only after the Oliver merger of 1929 were the tractor designs upgraded to vertical four-cylinder engines mounted lengthwise on the tractor frame.

OLIVER ROW-CROP TIPTOES IN

Oliver Chilled Plow engineers started work on a row-crop tractor after J. D. Oliver determined that Henry Ford did not plan to develop a row-crop design. By 1926, a prototype Oliver row-crop tractor was at work in Texas, based on a design by Oliver Chilled Plow engineer Herman Altgelt. He is credited with redesigning the wheels into the tiptoe configuration to handle the sticky gumbo at the Texas test site. Patent applications for the new row-crop tractor were filed in 1928, and during that year nine prototypes were tested. Each design had the name Oliver cast in the radiator, was painted gray with red wheels, and was designated either as Model A for the tricycle row-crop or Model B for the standard-tread machine.

The 1929 formation of the new Oliver Farm Equipment Sales Company changed everything, but the Oliver design did shape the new firm's entry into the row-crop tractor market. Oliver first used "floating"

Oliver 18/27
Oliver engineer Herman Altgelt designed the 18/27's tiptoe wheels to handle Texas gumbo soils. Subsequent Oliver row-crop tractors used two smaller front wheels instead of the large single wheel. Oliver's concept of large rear wheels eliminated the use of drop box gearing to gain under-axle crop clearance. The drive wheels were clamped to the splined rear axles, so they could easily be adjusted to the proper row width.

1929 Cockshutt Hart-Parr 18/28
Cockshutt sold Oliver Hart-Parr tractors painted the Canadian company's red and cream colors.

1929 Cockshutt Hart-Parr 18/28
Oliver manufactured several tractor models that were sold by Cockshutt Plow Company of Brantford, Ontario. This 1929 Cockshutt 18/28 was built in Charles City, Iowa, and its four-cylinder engine has a 4.125x5.25-inch bore and stroke. Cockshutt also sold other Oliver models, including the 28/44 standard, the early row-crop 18/28, and the streamlined 60 and 70 models until 1946, when it began manufacturing its own tractor models. Richard C. Brown of Harriston, Ontario, found and restored this beauty.

1930 Oliver Hart-Parr 28/44
The 1930 Oliver Hart-Parr 28/44 still carried both founding company names. The 4.75x6.25-inch four-cylinder engine in the 28/44 was used from 1930 until 1937, when the tractor evolved into the Model 90. Gradually, the Hart-Parr name disappeared from the Oliver tractors. Dave Preuhs, with help from his son Wayne, found and restored this early plow-pulling standard tractor near their home in Le Center, Minnesota.

adjustable rear wheels on the prototype South Bend tractor, and other tractor makers soon followed. New engines, drivetrains, and transmissions transformed the South Bend design into the Oliver row-crop machines produced starting in February 1930.

In 1930, the year following its founding, Oliver introduced its first row-crop tractor, a tricycle-type machine with a single front wheel. Skeleton steel rear wheels had a narrow vertical band to which twenty spade lugs were alternately attached left then right, giving it "power on tip-toe," according to Oliver ads. The large-diameter rear wheels were attached to the rear axles with a hub that could be loosened. That allowed the wheels to slide on the axle to the desired row spacing, an idea that was quickly used by both IH

and Deere on their second-generation row-crop tractors. On Oliver row-crops built from 1931 to 1937, the large, single front wheel was replaced with two close-spaced, smaller-diameter wheels.

Oliver's row-crop was powered with a four-cylinder engine, which was closely related in design to the engines in the company's new line of standard-tread tractors. The row-crop's transmission was also from the standards.

Oliver Hart-Parr standard tractors for 1930 included the 18/28 models, available in standard, western, rice field, or orchard versions, built until 1937. The larger 28/44 was built from 1930 to 1937 with either a four- or six-cylinder engine. It later became the Oliver Model 90.

1937 Oliver 70 Row-Crop
No, you aren't seeing things. This red Oliver was used in a color preference study at state and regional fairs to let farmers vote on the color they wanted on their Oliver tractors. Oliver's green color won. William Cover bought the red Oliver new after the 1937 study at the York (Pennsylvania) Interstate Fair. His grandson and grandson's wife, William K. and Melissa Cover, now own the tractor. Her father, Thomas Humbert, and brother Dale of McConnellsburg, Pennsylvania, did the sparkling restoration.

Right: **1935 Oliver Row-Crop 70 HC**
The Oliver 70 tractor's canvas sling seat addressed driver comfort.

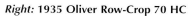

Oliver 70 Standard
The standard-tread version of the Oliver 70 brought the smooth power of a six-cylinder engine to even more farmers. The Model 70 engine had six 3.125x 4.375-inch cylinders and spun the crankshaft at 1,500 rpm to make 22.64 drawbar horsepower and 35.14 belt horsepower. Kerosene, as well as gasoline, engines were available in row-crop, standard, and orchard versions. John Clow of Centreville, Maryland, restored this steel-wheeled tractor. Electric lights and a starter were popular options.

1930s Oliver 70 advertisement

1939 Oliver 80 Row-Crop
Row-crop tractors began to come with more available power in the late 1930s. This 1939 Oliver 80 row-crop is a three-plow tractor with a four-cylinder gas engine. Its wide front axle is adjustable to fit crop row widths. Bill Meeker and his son Sam worked together to restore this vintage Oliver. They live near North Henderson, Illinois.

The Oliver Hart-Parr Row-Crop 70 HC, a streamlined, high-compression, gasoline-fueled six-cylinder row-crop on pneumatic tires, came out in October 1935. Its quiet, smooth-running engine attracted a lot of attention, and the tractor was immediately popular. Oliver sold 5,000 by February 1936, some 3,000 more than expected. That success proved that high-test gasoline was a viable tractor fuel. A companion tractor, the Row-Crop 70 KD, which could burn kerosene or distillate, was announced later that year. A starter and lights were available on both.

Oliver dropped the Hart-Parr name from its tractors in 1937 when it debuted an improved streamlined Model 70 row-crop.

Nebraska tests in 1940 showed 22.64 drawbar horsepower and 28.37 belt horsepower from the high-octane row-crop's six-cylinder 3.125x 4.625-inch engine running at 1,500 rpm. The 70 was also available in standard, industrial, and orchard versions, made from 1937 to 1948.

New in 1938 was the Model 80 row-crop tractor, a three-plow version of the 70. It, too, was available with either 70-octane or kerosene/distillate-burning engines. The KD 80 (kerosene-distillate) tractor pumped out 23.27 drawbar horsepower and 35.14 belt horsepower in its 1938 Nebraska tests. Its engine was a four-cylinder design with a 4.5x5.25-inch bore and stroke that ran at

1940 Oliver 80 Row-Crop Diesel
Diesel engines became available in the Oliver tractor line in 1940. A four-cylinder Buda-Lanova engine powers this Oliver 80 row-crop diesel. The Buda engine was eventually replaced by an Oliver diesel engine design. Oliver collector and professional tractor restorer Lyle Dumont of Sigourney, Iowa, located this tractor that was sold new in Stockton, California. The Oliver 80 Row-Crop was in production from 1937 to 1948.

1,200 rpm. In 1940, the Model 80 was offered in a diesel version with a Buda-Lanova engine. Oliver later made its own diesel engine. Styling on the Model 80 was apparently not a principal consideration. The tractor still carried the boxy look of the earlier tractors. Adjustable-tread-width axle front ends were available for row-crop conditions where the extra stability was needed.

Oliver standard-tread tractors for 1937 included an improved 28/44 called the Model 90. It had an electric starter, pressure lubrication, and a four-speed transmission. The Model 90 was replaced with the Model 99 in 1953, which stayed in production until 1957. The Oliver 80 Standard was the new designation for the old 18/28

Oliver Hart-Parr. Its high-octane gasoline engine produced 27.66 drawbar horsepower and 36.07 belt horsepower in 1940 Nebraska tests. It was powered by a 4.25x5.25-inch bore-and-stroke high-compression engine designed to run at 1,200 rpm.

MORE GEARS AND MORE GO

Highlighting Oliver's 1939 model introductions was a new small tractor, the Model 60 row-crop. The trimmed-down version of the popular 70 row-crop model was powered by a four-cylinder 3.3125x 3.5-inch engine. It weighed just 2,000 pounds on rubber tires. Initially available with a four-speed transmission, the Model

60 got a five-speed gearbox after production began and eventually came equipped with battery ignition and electric starting.

Oliver also soon enhanced its 70 model with a six-speed transmission and upgraded its standard tractors to four-speed gearboxes. By this time, all of the Oliver tractors had lights and starters.

CLETRAC CRAWLERS

Oliver added another dimension to its line in 1944 when it took over the Cleveland Tractor Company of Cleveland, Ohio, and added the Cletrac line of farm crawlers to Oliver's Meadow Green–painted tractors. Oliver then shortened its name to the Oliver Corporation.

Cleveland Tractor traced its history back to 1911, when Rollin H. White (famous for his steam auto and gasoline engines) and his brother Clarence developed a tracklayer tractor called the Cleveland Motor Plow. Unlike the Caterpillar tractors of its day, the Cleveland tractor used differential steering. Brake bands actuated by a steering wheel slowed one track and speeded up the other to turn the machine. Caterpillar tractors used two

1945 Oliver 60
Oliver introduced the Model 60 row-crop tractor in 1940, a smaller version of its famous Model 70. The 60 was available as a row-crop, standard, or industrial machine. The two-plow machine's power choices were a four-cylinder gas or distillate engine. Internal hydraulics, excellent visibility, and responsive steering made the 60 a pleasure to drive. Dr. Ferrel A. Moots of Kirksville, Missouri, found his father's 1945 Model 60 in Kansas. He bought it in 1991, returned it to Missouri, and restored it.

1945 Oliver 60 Industrial
The first of only 815 Oliver 60 Industrial tractors made, this 1945 model bears serial no. 46001. The Oliver 60 was made from 1940 through 1948, when it was replaced by a new Fleetline Series Model 66. Peter Petrovic of Duluth, Minnesota, found and restored the little tractor. Petrovic, a professional watchmaker, enjoyed his off-hours hobby and said it was a relief to make tractor adjustments in the evenings with a sledge hammer, rather than worrying that a sneeze might blow away delicate parts (as his day job required).

lever-operated steering clutches to slow or stop one track and let the other continue running to make its crawler turn.

Cleveland Motor Plow Company was incorporated in 1916 and soon produced its first tractor, the Model R with 10 drawbar horsepower. Early company ads touted the tractor as "Geared to the Ground." The company name changed to Cleveland Tractor Company in 1917, and the Cletrac trademark was adopted in 1918. The company's models included the 12/20 H of 1918, the 12/20 W of 1920, and the small 9/16 F of 1921. Some forty different Cletrac models were made between 1917 and 1944, when the company was sold to Oliver.

Cletrac Model 80 of 1933 was its first diesel. A smaller diesel, the Model 35, was available in 1934. Row-crop farming was addressed in the 1935 E model, which could be purchased in one of five tread widths ranging from 31 to 76 inches. The three- to four-plow E could cultivate four rows with a front-mounted cultivator. A more compact model, the HG, debuted in 1939. It, too, was available in different tread widths to accommodate row-cropping applications.

1939 Cletrac General GG
The Cletrac General GG was the only rubber-tired tractor made by the Cleveland Tractor Company, manufacturer of the tracklayer Cletracs. The General was announced in 1939. Two years later, the General tractor line was sold to B. F. Avery of Louisville, Kentucky, who developed the little tractor into three different versions. Oliver did not make the General GG model, but did make a small Cletrac row-crop crawler—the Model HG—using the same engine that was in the General. William G. Menke III of Mentor, Ohio, rebuilt this General after he bought it in Indiana in 1987.

The first and only Cletrac wheel-type tractor, the Model GG or General, was made and sold between 1938 and 1941. The General was a tricycle rubber-tired row-crop machine with a single front wheel. It was powered by a Hercules four-cylinder 3x4-inch engine. When tested at Nebraska, the GG produced 14.26 drawbar horsepower and 19.29 belt horsepower. It was later sold by the B. F. Avery Company of Louisville, Kentucky, and eventually ended up as part of the Minneapolis-Moline product line. Oliver didn't need the General because it had its own small row-crop tractor.

Oliver restyled the Cletrac HG crawler in 1947, painted it Meadow Green, and offered it for farm use in four tread widths of 31, 42, 60, and 68 inches. The HG's four-cylinder engine developed 18 drawbar horsepower and 22 belt horsepower.

A rubber-belted version of the HG, the HGR-42, was given a field trial in 1947, after several preproduction units were assembled. The gasoline tractor was basically a standard HG, but it incorporated a continuous belt-type "rubber band" instead of steel tracks.

***This page and next:* 1939 Cletrac E-62**
With a track spacing of 62 inches, this 1939 Cletrac E-62 crawler tractor qualified as a row-crop machine. Cleveland Tractor Company designed this crawler to front-mount a four-row cultivator to work four 34-inch rows of potatoes. It could also handle three to four plows for primary tillage. With an optional belt pulley, electric lights, and starter, the Cletrac E-62 had even more uses on the farm. Cletrac made many different models from its founding in 1916 to its sale to the Oliver Corporation in 1944. Collector Wilbur Lutz of Sinking Spring, Pennsylvania, uses barns on his farm to store his growing Cletrac and Oliver collections.

The innovative rubber-belt crawler was not a success. While test engineers found that more of the rubber track stayed on the ground over uneven terrain, it was difficult to keep properly tensioned. In fact, the rubber track stretched some when used, and the rubber compounds didn't hold up well. Work on the HGR-42 soon stopped, and it was forty years before the rubber-belt concept was used again. Then it was Caterpillar that introduced the concept on its Challenger farm crawlers in 1987.

LIVE PTO IN THE FLEETLINE FLEET

In 1948, a new line of Oliver tractors was announced to celebrate the centennial of John Nichols' founding of Nichols & Shepard, one of the Oliver founding firms. The new Fleetline tractors, complete with new grilles and sheet metal styling, were Models 66, 77, and 88. They were all equipped with independent, or live, PTO—

Cleveland Tractor Company advertisement

Above: 1950 Oliver Cletrac HG 68
Oliver green graces this 1950 Oliver Cletrac HG 68 crawler. The Oliver Cletrac HG model was powered with a Hercules four-cylinder engine that produced 22 drawbar horsepower and 27 belt horsepower from its 3.25x4-inch cylinders at 1,700 rpm. Numbers in its model designation refer to track widths available for different applications. Former Oliver dealer Robert H. Tallman of Harbeson, Delaware, bought this tractor back from a customer in 1985 and restored in when he began collecting Olivers. Production of Oliver Cletrac tractors ended in 1965.

Left: 1950 Oliver Cletrac HG 68
The 68 inches between the HG 68 tracks allowed it to straddle two 34-inch rows, a common width for potato crops.

1947 Oliver Cletrac HGR

Oliver decided to fit its HGR crawler with tracks made of continuous rubber belts in 1947. The crawler used a modified undercarriage with four track rollers and special track tensioning springs at the front idler. As many as 700 HGRs are said to have been made, but most were later recalled and converted back to the steel track undercarriage. This tractor has 42-inch track spacing for general uses. Caterpillar reintroduced the rubber belt crawler concept on its Challenger line models in the 1990s. John Deere also offers rubber belt tracks on some of its more powerful tractors. Tracklayer collector Harlan Thompson of Harper, Kansas, bought this HGR in pieces and had it restored and repainted in 1995.

1947 Oliver Cletrac HGR

Problems with the composition of the rubber track caused the Cletrac experiment to be cancelled.

1951 Cockshutt 30 Diesel
This 1951 Cockshutt Model 30 diesel is an import. The handy-sized row-crop has a four-cylinder Buda diesel engine that produced 28 horsepower on its drawbar and 32 on the belt in Nebraska tests. Cockshutt Farm Equipment of Canada made its own tractors at its Brantford, Ontario, plant from 1946 until 1962. Previously, the company sold Oliver-made tractors from 1930 until 1946. Cockshutt made its 30, 40, 20, and 50 models until 1957, when it upgraded them to the 500 Series. Jeff Gravert of Central City, Nebraska, restored this Cockshutt with a three-point hydraulic hitch and single front wheel.

1948 Co-op E3
Equivalent to a Cockshutt Model 30, this 1948 Co-op E3 was sold by The National Farm Machinery Cooperative, Inc. of Bellevue, Ohio. The E3 is powered with a 153-cubic-inch 3.0625x4.125 bore-and-stroke 1,650-rpm gas engine. Fittingly, Midland CO-OP, Inc. of Stilesville, Indiana, restored the tractor and uses it around its elevator. Dave Frayser at Midland painted the tractor in the burnt orange Co-op color.

1949 Co-op E3
This 1949 Co-op E3, more widely recognized as a Cockshutt Model 30 standard-tread in burnt-orange paint, was made by Canadian Cooperative Implement Limited (CCIL). Randy Bader of Edison, Ohio, bought this tractor in 1994 and restored it with the help of his brother Gary.

the first wheel-type tractors in the United States to be so equipped. The feature permitted a PTO-driven machine to run whether the tractor was in motion or at rest.

The one- to two-plow Model 66 came with three engine choices: a 129-cubic-inch high-compression gasoline or diesel engine; or a 145-cubic-inch lower-compression kerosene/distillate engine. Oliver also offered kerosene/distillate engines of larger displacement but lower compression in the new 77 and 88. The two- to three-plow Model 77 offered a six-cylinder 194-cubic-inch

engine in either gasoline or diesel versions. The three- to four-plow Model 88 came with a six-cylinder 231-cubic-inch engine as either gas or diesel. Standard-tread and adjustable wide-front versions were offered in the 66, 77, and 88 models. Well-shielded orchard models were also available.

A new Oliver compact utility-type tractor, the Super 55, was announced in 1954. It was available with a gas or diesel engine and featured a three-point hitch with draft control. The trend toward compact utility-type tractors began with the Ford-Ferguson 9N of

1957 Cockshutt Golden Arrow

Cockshutt's 1957 Golden Arrow was a special show tractor made in limited numbers to present the draft-sensing three-point hitch offered on forthcoming 540 and 550 models. The Golden Arrow had the engine of the Cockshutt Model 35 with the transmission and differential assembly of the 550. The tractor's deluxe styling reversed the paint colors used on earlier models with cream as the prominent color and red as the trim color. This is a western tractor with standard gear and deep crown fenders. Cockshutt specialist Gary Bader of Marion, Ohio, restored the tractor.

1939. In this type, the operator sat low, slightly forward of the rear axle, and straddled the transmission case on a seat mounted above the differential. The Super 55 gas engine delivered 29.60 drawbar horsepower and 34.39 belt horsepower. The diesel version of the Super 55 ran at 33.71 belt horsepower.

The Super Series also included the Super 66, Super 77, Super 88, and Super 99, all with more power. The Supers were made from 1954 to 1958. The Super 44 was a new, smaller, utility-type tractor with the engine and drivetrain offset to the left to permit better vision for cultivating with the one-row

1961 Cockshutt 580 Super Diesel
This 1961 prototype Cockshutt 580 Super diesel is the only one of its kind. It was never manufactured. The big standard-tread tractor has a six-cylinder 100-horsepower Perkins diesel engine with plenty of power to pull the attached five-bottom plow. Cockshutt was bought by White Motor Company in January 1962, and work on this new model was stopped. White already owned Oliver and went on to buy Minneapolis-Moline in 1963. Oliver again supplied its tractor models to Cockshutt for Canadian sales from the mid-1960s until 1977 when the Canadian name disappeared. McComas Albaugh, Paul Summers Jr., and Clayton Lenhart shared this big Cockshutt kept at Union Bridge, Maryland.

machine. Its power came from a 139.6-cubic-inch four-cylinder Continental L-head engine.

The Super 99 diesel of 1957 replaced the older 99 model and had two diesel engine offerings. Farmers could opt for the six-cylinder Oliver diesel or a three-cylinder, two-cycle General Motors diesel. Optional equipment on the Super 99 included a torque converter transmission and an all-weather cab. Nebraska test No. 556 recorded 73.31 drawbar horsepower and 78.74 belt horsepower from the blower-equipped GM two-cycle diesel version of the Super 99. Oliver now had a new strong contender in the horsepower race.

In 1958, the Oliver tractor line was given new radiator grille styling with bolder horizontal stripes. The Meadow Green machines were accented by Clover White wheels and grilles. Model numbers became 440, 550, 660, 770, 880, and 990. Also available were Models 950 and 995. More power and improved power transmissions were featured on all the models. Full-time power steering became standard. The Oliver 995 GM Lugmatic diesel had an automatic torque converter and produced 85.37 PTO horsepower and 71.44 drawbar horsepower.

Bigger changes for Oliver came in 1960, after introductions of new models with the square-section egg-crate grilles. New

1953 Oliver 77 with wide front
Bob Johnston of Findlay, Ohio, added a chrome exhaust stack when he and his son Scott restored this 1953 Oliver 77 with a wide, adjustable front end. Only about 10 percent of the row-crop Model 77s came with the wide front. Bob converted his tricycle to the rare wide-stance front after he found one for sale in Creston, Ohio.

were the Model 500, with a David Brown engine, and the 1800 and 1900 models.

OLIVER SWALLOWED WHOLE

In November 1960, the White Motor Company of Cleveland, Ohio, bought Oliver. White followed that with the purchase of Cockshutt in 1962 and Minneapolis-Moline in 1963. The products from these acquisitions were gradually merged in the intervening years, and in 1969, the three divisions became White Farm Equipment, a wholly-owned subsidiary of the White Motor Corporation.

The last Oliver tractor in green paint that still bore the Oliver name was manufactured in 1976. It was an Oliver Model 2255—

1953 Oliver 77 Row-Crop
A yellow grille with horizontal slots gave the Oliver Fleetline tractors a bold new look.

1949 Oliver 77 with No. 4 Corn Picker

The Oliver 77 row-crop tractor could not only cultivate corn, it could harvest it, too. The two-row mounted corn picker here is an Oliver No. 4 model. The picker snapped the corn ears off of the stalks, elevated them to a husking bed that removed the husks, and then elevated the shucked ears and dropped them into a trailing wagon. After harvest, the picker was removed from the tractor, freeing it up for fall and spring tillage uses. Bruce B. Youde of Polk City, Iowa, restored this 1949 tractor–corn picker combination for his growing collection of Olivers.

1949 Oliver 77 with No. 4 Corn Picker

Tractor-mounted corn pickers took the hard work out of corn harvest. The corn picker was later replaced by the combine. Combines now shell the corn off the cob in the field.

Above: **1951 Oliver 77 Orchard**

The 77 Orchard's turf tires gave roots and soil-conserving interplantings a break. The tractor's operator and tractor controls are protected from limb strikes by shielding.

Right: **1951 Oliver 77 Orchard**

The flowing fenders and shields on this 1951 Oliver 77 Orchard are for utility rather than streamlined beauty. Even the headlights are mounted in a low position so they don't catch on tree branches. Oliver's Model 77 was a smooth-running machine powered by a six-cylinder 194-cubic-inch engine in either gas or diesel versions. Live PTO was a welcome feature on the tractor. Verlan Heberer of Belleville, Illinois, collected and restored this beauty.

1950s Oliver 77 and 88 advertisement

1953 Oliver 99 B

The new Oliver heavy-duty tractor introduced in 1953 was the Model 99 B standard-tread machine. The Model 99 had a four-speed transmission and could pull four to five plow bottoms with its powerful six-cylinder gas overhead valve engine. Less than fifty of the gas-burning 99 models were built in 1953. Oliver collector-restorer Edward J. Schulte of Manchester, Iowa, restored this big tractor. Its Oliver stablemates were the smaller 88, 77, and 66 models. They were available in standard, row-crop, and high-clearance versions.

a big tractor sporting a 636-cubic-inch Caterpillar diesel engine. It soon became one of the early White company's silver-and-charcoal Field Boss models.

In 1981, TIC Investment Corporation of Dallas, Texas, bought White Farm Equipment and continued its operation as WFE. Hard economic times caused WFE to file for bankruptcy in 1985. Allied Products Corporation of Chicago subsequently bought parts of WFE, including the tractor factory in Charles City, Iowa, and all tooling and inventory. By 1987, Allied had combined White Farm Equipment with its New Idea company into a new division called White-New Idea. In 1993, AGCO bought the White-New Idea implement line.

1957 Oliver Super 99 GM

A big, turbocharged General Motors three-cylinder two-cycle diesel engine with torque converter put the go into the 1957 Oliver Super 99 GM tractor. The tractor's 213-cubic-inch powerplant produced 73.31 drawbar horsepower and 78.74 belt horsepower at 1,675 rpm in its 1957 Nebraska tests. The Super 99 GM had a six-speed transmission and could run from 2.6 to 13.7 miles per hour. It weighed 10,155 pounds when tested. Derived from the standard Model 99, the Super 99 came equipped with either a six-cylinder engine or one of two diesel engines. The diesel options were the Oliver six-cylinder diesel or the GM engine, like the one in the photograph. Owner Bruce Miller of LaMoille, Illinois, worked the tractor on his farm before it was restored.

Chapter Eleven
Orphans and Others

1919 Avery 14/28
The Avery Company of Peoria, Illinois, a steam engine and threshing machine maker, was a prolific developer of gas farm traction engines as steam engines fell out of favor. One of Avery's many offerings was this "small" 1919 Avery 14/28 four-cylinder horizontally opposed "gas oil" tractor. Its 4.626x7-inch cylinder engine ran at 700 to 800 rpm. A later upgrade kit promised 20/35 performance with larger sleeves, pistons, and rings—for just $89. Kenneth L. Lage of Wilton, Iowa, owns this interesting piece of history with remarkable pin striping.

From the hundreds of manufacturers and inventors active in the farm tractor business between 1900 and 1960, less than a dozen survived into the 1960s. Some makers were fortunate enough to be bought out or merged into other companies. Some dabbled in making tractors after World War II to meet demand, but then stopped once the tractor market cooled down. Others, once successful and famous as steam traction engine and threshing machine makers, didn't transition to gasoline power fast enough to compete and continue. Now the relics of these failed efforts are known among collectors as orphans.

By about 1915, swirling changes in farm power swept many makers into oblivion. The writing was on the wall: the steam engine was doomed by the new gas-powered tractor. By 1925, steam machines were old news and gas tractors were all the rage. They were becoming agile, all-around machines capable of handling almost any farm power task.

Recurring economic recessions and depressions also took their toll. Some makers with successful products fell on hard times and had to throw in the towel when receivables soared. A combination of liberal credit terms and hard times on the farm did in many makers.

In this chapter, some of the tractors that live on from these doomed makers are highlighted. They are restored examples that show the types of mechanization that revolutionized farming over the years.

AVERY COMPANY

The Avery Company of Peoria, Illinois, was a well-known power in farm steam engines and threshing equipment. Its famous undermounted steam engine faced formidable competition from the emerging gasoline-engine machines, so Avery began its defense by building an innovative gas tractor-truck in 1909. The new machine looked more like a truck than a tractor, but it was designed to pull as well as haul. Round wooden plugs made up the tread on the model's wheels. It was discontinued in 1914.

About the same time, Avery built a huge, one-cylinder, 12x18-inch gas engine on a tractor with a potential output of 65 horsepower.

Demonstrated at the 1910 Winnipeg trials, the tractor was withdrawn halfway through the trials because of its poor showing.

Following that shaky entry into the tractor market, Avery moved forcefully into the gasoline tractor business, introducing more than one model each year between 1911 and 1924. Avery spawned a stable full of light cultivating tractors between 1916 and the early 1920s to increase its market share. In moving up from a

1938 Avery Ro-Trak
After a second bankruptcy in 1931, Avery returned with a try at a combination row-crop and standard-tread tractor. The 1938 Avery Ro-Trak front wheels could be set at standard tread widths or narrowed together as a tricycle type. The two-plow tractor was powered with a six-cylinder Hercules engine. The Ro-Trak concept failed, and Avery was soon out of business. James Layton of Federalsburg, Maryland, collected this rare machine.

Avery Ro-Track advertisement

Prominent and respected steam engine maker A. D. Baker Company of Swanton, Ohio, was one pioneer who sought to keep farm steam engines competitive with gasoline tractors by designing lightweight and more efficient steam engines. The Baker Company built steam engines from 1898 until about 1925, when it at last acknowledged the end of steam and changed over to gas-powered machines.

The first Baker gas tractor was its 22/40. Using mostly off-the-shelf components from other manufacturers, Baker had its husky 25/50 model ready by late 1927. When tested at Nebraska in 1929, the 25/50 delivered a surprising 43 horsepower on the drawbar and 67 on the belt. So, unlike many other tractors tested, the big Baker was redesignated as a bigger 43/67, compared with the original designation of 25/50. Baker tractor production ended during World War II.

B. F. AVERY COMPANY

Making tillage tools for the southern farm market was an early strength of the B. F. Avery Company of Louisville, Kentucky. Benjamin Franklin Avery started making farm tools in about 1825 in his Clarksville, Virginia, blacksmith shop. By 1845, Avery's plow factory had moved to Louisville. At its incorporation in 1877, the company was capitalized at $1.5 million. Through growth of the company, Avery began to distribute its products nationally and internationally in 1939.

Avery's first tractor—a three-wheeled machine called the Louisville Motor Plow—was produced in 1915. The 2.5-ton tractor was powered by a 20-horsepower engine. Its two plow bottoms were mounted on the tractor frame and could be removed for other work. The tractor was built for only a very short time.

Avery acquired the Cletrac General GG row-crop tractor from Oliver in 1944. Oliver didn't need the rubber-tired machine in its tractor line after it meshed the newly purchased Cleveland Tractor Company crawler tractors into its wheel tractor offerings. Avery then sold different versions of an improved GG tractor under its own name and apparently supplied the tractor to other short-line retailers. In 1951, Minneapolis-Moline bought the B. F. Avery Company and began selling the former General tractor in M-M Prairie Gold paint.

one-row four-cylinder 5/10 cultivating tractor to a two-row unit, Avery boosted the tractor's engine to six cylinders on the Model C. By 1919, Avery was making eight different gas-powered tractor models, ranging in power from a 5/10 lightweight unit up to a giant 40/80 tractor (later redesignated as the 45/65 following its Nebraska tests). Avery also developed and tested a four-cylinder 15/25 Track Runner, a front-wheel steered crawler, in 1923. In 1924, Avery filed for bankruptcy.

Reorganized as the Avery Power Machinery Company, the new company succeeded in keeping the wolf away from its doors until 1931, when it again went under. After another reorganization, Avery produced the Ro-Trak tractor in the late 1930s, an innovative effort that promised the benefits of a row-crop machine and the stability of a standard-tread tractor. That machine was Avery's last.

Above: 1950 B. F. Avery A

This 1950 B. F. Avery Model A tractor is not related to the Avery Company of Peoria. B. F. Avery of Louisville, Kentucky, bought the rights to the small Cletrac General Model GG tricycle row-crop in 1941 from its Cleveland, Ohio, maker. B. F. Avery used the same girder frame as the Cletrac General, but added optional wide front and two-wheel tricycle front ends to the single wheel front. This B. F. Avery A is powered with a four-cylinder Hercules 3x4-inch engine that makes about 14.26 drawbar horsepower and 19.29 belt horsepower. Billie and Russell Miner of Pleasant Plains, Illinois, restored this tractor.

Left: 1950 B. F. Avery A

The Model A's wide front aids in the tractor's versatility.

BATES STEEL MULE

Joliet Oil Tractor Company of Joliet, Illinois, and the Bates Tractor Company of Lansing, Michigan, merged in 1919 to become the Bates Machine and Tractor Company, eventually producing the Bates Steel Mule. The crawler tractor with front wheels was similar in design to other small tracklayers of the era.

By 1924, Bates had five models in production. Its industrial Model 25 and Model 50 of 1924 tracklayers no longer had wheels in front. They were followed with Bates' Model 80 and Model 40 diesel in 1929 and 1937. Bates tractor production ended in 1937.

EAGLE MANUFACTURING

Eagle Manufacturing Company of Appleton, Wisconsin, made a variety of gasoline-powered farm tractors, beginning with a 1906 machine driven by a two-cylinder horizontally opposed engine. A 1911 model of 56 horsepower employed a four-cylinder engine. By 1916, Eagle was selling 8/16- and 16/30-sized tractors, and it progressed through the 1920s with a series of two-cylinder machines.

In 1930, Eagle jumped from equipping its tractors with horizontal two-cylinder engines to using vertical six-cylinders to power its 6A Eagles. The 6A was a three- to four-plow tractor with

1920 Baker 25/50
Respected steam engine maker A. D. Baker of Swanton, Ohio, switched quickly to gas tractors around 1927 when steam fizzled its last. Baker was so conservative in its power ratings that this 25/50 tractor tested out at 43 drawbar horsepower and 67 belt horsepower at Nebraska. It then became the 43/67 model. Its four-cylinder gas engine had a 5.5x7-inch bore and stroke. A previous owner thought enough of this machine in 1938 to return it to the Baker factory for conversion to rubber tires. German exchange student Rudiger Herbst helped Darius Harms of St. Joseph, Illinois, restore the Baker, pin stripes and all.

1923 Bates F Steel Mule
Wheels and steel tracks were combined on this 1923 18/25 Model F Bates Steel Mule. The front wheels steered the machine as the tracks pushed. Bates made this model from 1921 to 1937 with a LeRoi 4.25x6-inch four-cylinder engine. Bates made a full-fledged crawler, the Industrial 25, from 1924 to 1928, apparently learning that the front wheels were not needed to steer a crawler design. Bates also offered a Model H tractor that was of a conventional wheel design.

1936 Co-op No. 3
A six-cylinder Chrysler industrial engine propelled this 1936 Co-op No. 3 standard-tread tractor with 29 drawbar horsepower. Farmers Union Cooperative sold the machines made by Co-Operative Manufacturing of Battle Creek, Michigan. This tractor's five-speed transmission allowed it to move up to 35 miles per hour in road gear. The three-plow tractor was the largest of three models made. Duplex Machinery Company, also of Battle Creek, offered the three models in 1937 and 1938. Richard Bean of Chaseburg, Wisconsin, grew up wanting one of these tractors and later found and restored this one.

1932 Eagle 6A
Eagle Manufacturing of Appleton, Wisconsin, made this 1932 Eagle Model 6A standard tractor in 1932. It has a six-cylinder 4x4.75-inch Waukesha engine that produced 29.52 drawbar horsepower and 40.36 belt horsepower. Eagle began making tractors in 1906. They were mostly horizontal two-cylinder machines until 1930, when the company developed and marketed the modern Eagle 6A. Eagle next added a 6B Universal row-crop and a 6C utility model powered by a Hercules six-cylinder engine with a 3.35x4.125-inch bore and stroke. Eagle tractor production didn't resume after World War II. Robert Brennan of Emmett, Michigan, collected and restored this Eagle he first spied in 1980.

a Waukesha 4x4.725-inch engine, giving it a 22/37 rating. In following years, the Eagle line expanded to include a Model 6B row-crop and a 6C utility version, both powered with Hercules sixes of 3.25x4.125-inch bore and stroke. The uncertainties of World War II doomed the Eagle tractors.

SEARS ROEBUCK SELLS GRAHAM-BRADLEY TRACTORS

With lines and speed suggestive of its automotive heritage, the six-cylinder Graham-Bradley 20/30, built by Graham-Paige Motors Corporation of Detroit, Michigan, was a hit when it was introduced in 1938. Initially, it was marketed through mail-order giant

Sears Roebuck & Company of Chicago, Illinois, and was available at their retail stores or could be ordered from their huge catalog.

The Graham-Paige company had built a reputation on its high-performance vehicles and race cars. With a high road gear capable of speeds up to 20 miles an hour, the racy tractor took advantage of the technology of the times and came standard from the factory with starter and lights, rubber tires, PTO, and hydraulic lift.

Both row-crop and standard-tread versions were built during the relatively short life of the Graham-Bradley. By 1941, production had ceased. After World War II, plans were announced to make the tractor again. Those were shelved when Graham-Paige

President Joseph Fraser merged the company with Henry J. Kaiser's interests and made the postwar automobiles—the Fraser, the Kaiser, and the smaller car, the Henry J.

GM GETS IN THE TRACTOR BUSINESS

In an effort to compete with Henry Ford's announced intention to build a lightweight tractor, General Motors of Pontiac, Michigan, researched many firms and, in 1918, bought the Samson Tractor Works of Stockton, California, manufacturer of the Samson Sieve-Grip tractor. GM subsequently bought a manufacturing plant at Janesville, Wisconsin, and expanded it to produce Samson tractors, specifically designed to compete with the recently introduced Fordson.

The Samson Sieve-Grip model was re-engined with a GMC truck motor and introduced in 1918. But priced at $1,750, it couldn't compete with the Fordson.

The Samson Model M was a Fordson look-alike that was designed and positioned to compete with the Fordson. It came out in late 1918, was priced at $650, and was apparently on target.

GMC announced another tractor, the Model D Iron Horse, at about the time as the Model M. GMC bought the Iron Horse design, which was originally called the Jim Dandy motor cultivator. The small four-wheel-drive machine worked something like today's skid-steer loaders. Belts with idler tensioners served as clutches for each side of the machine. With both belts tightened, the machine moved straight ahead. Weighted levers were lifted to loosen both drive belts to stop.

The tractor could be operated from its seat or driven like a team. Reins attached through the pulleys mounted above the levers allowed the machine to be controlled by an operator walking behind the machine. A tug to the left rein turned the machine

1916 Galloway Farmobile
This 1916 Farmobile, collected and restored by Kenny Kass of Dunkerton, Iowa, is a rare machine. Waterloo, Iowa, entrepreneur William Galloway made this tractor for just three years—from 1916 to 1919. The Farmobile was rated as a 12/20 tractor and was powered with an inline four-cylinder 4.5x5-inch engine tucked between its rear drive wheels. Its transmission was mounted in front of the engine and drove chains to sprockets on the drive wheels. Galloway went bankrupt in 1920, but recovered and by 1927 was making short-line farm equipment in Waterloo as William Galloway & Sons.

1949 Gibson I

Many new tractor makers entered the U. S. market after World War II as domestic producers scrambled to meet the pent-up demand. The Gibson Corporation of Longmont, Colorado, offered this 1949 Gibson Model I three-plow tricycle to the hungry market. Made mostly from off-the-shelf parts, the Gibson I harnessed a 40-horsepower inline six from Hercules as it power source. Gibson built smaller and larger tractors until the postwar demand was met. Collector Don Neil of Canton, Missouri, collected this 1949 Gibson I.

that way; a tug to right and the machine turned right. Pull harder on the reins and the tractor reversed, as long as the reins were held back.

Although the Iron Horse was attractively priced at $450, the motor cultivator, of any make, was never a roaring success. Although the Samson M was marginally profitable, the entire tractor venture was not, and GM sustained heavy losses in the new enterprise. The post–World War I economy was poor, and competition was fierce. GM dropped farm tractors like a hot potato in 1922, and has not competed in that arena since.

HUBER MANUFACTURING

With a farm equipment history that traced back to the Civil War, the Huber Manufacturing Company of Marion, Ohio, was already a veteran steam engine and thresher manufacturer in 1898. That year it bought the patent rights to the Van Duzen gasoline engine—the engine that had powered John Froehlich's famous 1892 tractor. Huber made thirty tractors with the single-cylinder Van Duzen engine, but apparently took a hiatus from tractor making after that.

In 1911, Huber resumed making tractors, these powered with two-cylinder horizontally opposed engines. Huber's Farmer's Tractor was expanded to two models by 1912. Also joining the model lineup was a 30/60 four-cylinder heavyweight. Cross-mounted four-cylinder engines showed up in 1917 on the Huber Light Four of 12/25 horsepower. It was a popular machine and was built until 1928. A Super Four of 14/30 rating was announced in 1921. By 1925, its final year of production, the Super Four had grown to an 18/36 rating. An enclosed drivetrain with its cross-mount four was a feature of the Master Four of 1922. It was rated at 25/50. As modern as it was, the Master Four was replaced shortly with more modern designs.

Vertical four-cylinder inline engines marked the series of Huber tractors launched in 1926. The Super Four 18/36 had a Stearns overhead-valve engine that was upgraded to a 21/39 rating in 1929. The Huber 20/40 of 1927 also was powered by a Stearns OHV motor. Its rating was upped to 32/45 after its 1929 Nebraska test. The Model HK stayed in production until World War II. The bigger 25/50 of 1927 got its rating bumped up after its Nebraska

The roster of orphan tractors is long, filled with makes and models well known at their time, but now largely forgotten. Some include the following:

Buffalo-Pitts Company of Buffalo, New York, did major pioneering work on threshing machines from about 1834, and produced threshers and steam and gas tractors until the early 1920s. The company's thresher design was licensed to many American makers.

Bullock Tractor Company of Chicago, Illinois, started in 1913, making its Creeping Grip crawler tractors from ideas and patents from the bankrupt **Western Implement & Motor Company** of Davenport, Iowa.

Early tractor designer Albert O. Espe of Crookston, Minnesota, made his 1907 tractor at **C.O.D. Tractor Company** in Crookston. Espe's designs also were used by **Universal Iron Works** of Stillwater, Minnesota, for its two-cylinder Universal—a machine that later became the Rumely Gas Pull. Espe later designed Avery gas tractors and was once on that firm's payroll.

Electric Wheel Company of Quincy, Illinois, a manufacturer of metal wheels, experimented with tractors beginning in 1904, and in 1908 offered farmers a traction truck that could be converted to a farm tractor with engine mounting. By 1912, Electric Wheel was selling a 15/30 Model O All-Purpose Tractor. Other models followed, and the firm continued its line of tractors until about 1930. Electric Wheel became a part of Firestone Tire & Rubber Company in 1957.

Fairbanks, Morse & Company of Chicago, famous maker of weighing scales, added gas engines to its product line in the late 1800s. By 1910, the company entered the gas tractor market with machines made in Beloit, Wisconsin. It quit making its own tractors in 1914, but sold tractors made by other firms under the Fair-Mor name until 1918.

Gaar, Scott & Company of Richmond, Indiana, was another respected maker of steam traction engines, threshers, and sawmills. Its huge 40/70 gasoline tractor became the Rumely TigerPull when the M. Rumely Company bought out Gaar-Scott in 1911.

Gibson Corporation was one of several new companies that arose shortly after World War II to build tractors to meet industry demands. Gibson made several tractor models, ranging from garden tractor to three-plow size, but after a few years production of these was canceled.

Also in the post–World War II scramble to supply tractors to a hungry U.S. and world market, **Intercontinental** of Texas came out with a two-plow Model C-26 row-crop tractor in 1948. It was designed to compete with the popular Farmall Model H and featured hydraulics and electric lights and starter. Very few were sold in the United States, although some 2,000 were made for export sales. Intercontinental sold some tractors to the military as late as 1954.

Keck-Gonnerman Company of Mount Vernon, Indiana, made its first gas tractor in 1917, a two-cylinder 12/24. More standard-type models continued through the 1920s with four-cylinder inline engine models ranging from 18/35 up to a 30/60 Kay Gee Model N. Keck-Gonnerman stopped tractor production during World War II and didn't resume manufacture after the war.

Flour City tractors came from **Kinnard-Haines Company** of Minneapolis from 1889 to 1929, when the surviving company, Kinnard & Sons, was bought and tractor production stopped. The Flour City machines were mostly big four-cylinder inline tractors in power sizes to 40/70.

Leader Tractor built tractors between 1937 and 1948, including Models B and D, which were red-painted small tractors. A larger Leader Model A was built in a dozen units in 1944. There was an earlier Leader tractor made in Des Moines, Iowa, around 1918, but it was apparently not connected with the Chagrin Falls tractor maker.

The historic washing machine maker, the **Maytag Company** of Newton, Iowa, made a farm tractor briefly in 1916. The 12/25 tractor was produced for a short time during a period when farm implements were in Maytag's product line.

Catalog merchandiser **Montgomery Ward** of Chicago sold Wards tractors in 1950, built by Harry A. Lowther Company of Shelbyville, Indiana. Lowther's tractor used a fluid coupler from a Chrysler transmission and was called the Custom Gyrol Fluid Drive.

The **Otto Gas Engines Works Company**, a successor company to Nicolaus A. Otto, father of the four-cycle gas engine, made gasoline tractors in Philadelphia from 1896 until 1904. The company's tractors didn't evolve beyond the single-cylinder stage.

Seaforth, Ontario, was the home of **Robert Bell Engine & Thresher Company**, which sold and serviced the 22/40 Imperial Super-Drive. Its components were apparently made by the Illinois Tractor Company of Bloomington, Illinois. Illinois first offered its Illinois Super-Drive in 1919. As the Illinois Silo Company, the company began making light cultivating tractors in 1916. Its first larger tractor was the Illinois 12/30 of 1918.

Continued interest in the efficiency of the diesel engine prompted **R. H. Sheppard** of Hanover, Pennsylvania, to develop four farm-sized diesel tractors in 1949. The SD-1, SD-2, SD-3, and SD-4 offered one-, two-, three-, and four-cylinder diesel engines designed to pull one-, two-, three-, and four-bottom plows. The company also offered kits to install its engines in other makes of tractors. Sheppard diesel tractor production continued until 1956. Today, the Sheppard Company makes power-steering systems for trucks.

Of the tractor-plow type, the **Square Turn** tractor could turn in its own length. It existed under different ownerships between 1916 and the end of its production in 1925, when its assets were sold for $15,000 at a sheriff's sale. The tractor used machine power to lift its three-bottom plow.

L. M. Turner gained recognition for the Simplicity gasoline engines he built in Port Washington, Wisconsin. In 1915, his firm designed two sizes of farm tractors: the 12/20 and 14/24. Sales were initially good, but after World War I, the **Turner Manufacturing Company** failed and was liquidated in 1920.

Former Turner employee William J. Niederkorn and former Turner partner Francis Bloodgood bought the remaining assets of the bankrupt firm and in 1921 began the **Simplicity Manufacturing Company**. The company then produced cylinder boring machines. In 1937, Montgomery Ward and Company (then a mail-order business based in Chicago) had Simplicity make a small garden tractor. Simplicity was soon making a complete lawn and garden equipment line with its own national dealership network. Simplicity started building a riding garden tractor in 1939, and by 1958 it was making riding lawnmowers.

John Deere's grandson, Stephen H. Velie, began the **Velie Carriage Company** in 1902. That Moline, Illinois, firm was followed by several Velie companies, including the Velie Motor Vehicle Company that was organized in 1908 to make autos, and the Velie Engineering Company that was formed in 1911 to specialize in trucks. The Velie firms were combined in 1916 as Velie Motors Corporation, which built a well-engineered tractor called the Velie Biltwel beginning in late 1916. Using some of its automotive experience, Velie put a four-cylinder inline engine in the tractor. The Biltwel was a 12/24 standard-tread tractor with automotive-type steering. It weighed about 4,500 pounds. Improved in 1917 with even more automotive influence, the Biltwel was discontinued in 1920 in the face of stiff competition from cheaper tractors, including the Fordson. The Velie Biltwel was then priced at about $1,750.

Following a couple of unsuccessful ventures in the automobile business, the **William Galloway Company** of Waterloo built a 1916 Farmobile tractor in Waterloo, Iowa. Following bankruptcy three years later, the company reorganized as William Galloway & Sons and made short-line farm implements for many years, but without the Farmobile.

Yuba Ball Tread crawler tractors up to 40/70 ratings were made by the **Yuba Manufacturing Company** of Marysville, California, from designs originating with the Ball Tread Company of Detroit, Michigan. The Ball Tread tractor traveled on tracks running on large steel balls in a "race" under the tracks—similar in concept to a ball bearing. The design aimed at reducing friction inherent in crawler mechanisms. The tractor used a large single wheel in front to steer the machine. The Yuba Construction Company bought the Ball Tread Company and moved it to California in 1914. In 1918, the company became Yuba Manufacturing Company. No tractors were apparently made by the firm after 1931.

1928 Huber 25/50

Huber Manufacturing Company of Marion, Ohio, traced its origins back to Civil War days as a maker of farm equipment, including steam engines and threshing machines. This 1928 Huber 25/50 was powered with a 618-cubic-inch Stearns four-cylinder engine that demonstrated 50 drawbar horsepower and 69 horsepower on the belt when tested. It was a great plow and belt tractor. It weighed more than 9,000 pounds. Huber quit making farm tractors during World War II but continues to manufacture industrial equipment. Don and Marty Huber of Moline, Illinois, restored this big Huber to its original World War I army green color.

Huber Streamlined Model B advertisement

tests showed 50/69-horsepower performance. The tractor re-rated conservatively, though, as a 40/62 tractor. Its Stearns motor was of 5.5x6.5-inch bore and stroke.

Huber addressed its need for a row-crop tractor with the standard-tread 1931 Modern Farmer, which was equipped with an arched front axle and drop-box gearing on the rear axles to give it more crop clearance. A Light Four model of 1929 was a 20/36-rated tractor tested at Nebraska in 1929. Unlike the Super Fours and their Stearns powerplants, the Light Four was equipped with a Waukesha 4.75x 6.25-inch four-cylinder engine with a Ricardo head that promised prolonged fuel turbulence in the cylinder for better power and performance.

A Modern Farmer Model L (standard) and LC (tricycle row-crop) were announced in 1937 and were produced until World War II. They were of about 27/43 rating on rubber. Their power came from a 4.5x5.5-inch bore-and-stroke Waukesha VIK engine.

Both were available with electric lights and starters. The Huber Model B of 1937 was a streamlined row-crop of tricycle configuration powered with a Buda engine of 3.125x4.5-inch cylinder dimensions. An orchard version was also available. Huber tractor production ended in 1941, but the company continues to produce construction equipment.

PIONEER OF THE PRAIRIES

Pioneer began making tractors in Winona, Minnesota, in 1910 with the introduction of the Pioneer 30. The big machine stayed in the Pioneer line until 1927. The Pioneer 30 was driven by a four-cylinder

Graham-Bradley Tractor advertisement

1938 Graham-Bradley
The automotive influence of Graham-Paige of Detroit, Michigan, shows in the styling of this slick 1938 Graham-Bradley tractor. Initially sold through Sears & Roebuck retail stores, it was also promoted in the firm's mail-order catalogs. "The Graham-Bradley Tractor is modern farm power at its very best. Built by Graham . . . equipped by Bradley . . . guaranteed by Sears," the ads claimed. Graham's own six-cylinder 3.25x 4.375-inch engine powered the machine and produced its 20 drawbar horsepower and 28 belt horsepower. Electric lights and a starter, rubber tires, PTO, and a hydraulic lift were standard features of the tractor.

1922 Imperial Super Drive
Assembled in Seaforth, Ontario, by Robert Bell Engine & Thresher Company, the 1922 Imperial Super Drive is basically the Illinois Super Drive tractor under its Canadian trade name. The 22/40 tractor weighed 6,200 pounds and was powered with a Climax engine. Illinois Tractor Company of Bloomington, Illinois, made light cultivator tractors from 1916 to 1918 and added its Super Drive tractor in 1919. This tractor was bought new by Ron MacGregor's father in 1922 and has been on the family farm near Kippen, Ontario, since.

horizontally opposed 7x8-inch engine working through a three-speed transmission, giving it speeds up to six miles per hour.

A small Model 15 was also available, as was a gargantuan 45/90 with a six-cylinder engine and 9-foot-diameter drive wheels. In 1916, a 15/30 Pony Pioneer with only one drive wheel was offered. The company's 1917 Pioneer Special, also of a 15/30 rating, had its weight pared back to 8,500 pounds—compared with 24,000 pounds for the Model 30.

A more conventional tractor replaced the Special in 1919, an 18/36-rated machine of only 6,000 pounds. It retained the four-cylinder horizontally opposed engine design used in the earlier tractors. After 1927, Pioneer models didn't appear in tractor listings.

RUSSELL & COMPANY BUILDS GIANTS

Adapting a three-wheel English design, Russell & Company of Massillon, Ohio, began its transition from building steam engines and threshing machines to making gas tractors in 1909. The Russell American 20/40 was built from 1909 until 1914. A three-wheel 30/60 with a crossmounted four-cylinder engine was introduced in 1911 and later was called the Giant. In 1913, the Giant became a four-wheel tractor rated at 40/80. It was built between 1914 and 1921, but after Nebraska testing the tractor became the 30/60 and stayed in production until 1927.

Russell made lighter tractors beginning in 1915 with its 12/24 Russell Jr. The later tractors were of standard-tread design with

1948 Intercontinental C-26
One of the many new post-war tractor companies in the United States, Intercontinental Manufacturing of Garland, Texas, made its Model C-26 for domestic and export sales. Several thousand Intercontinental tractors were sold in Cuba, Mexico, India, and Argentina during the five or six years they were made. This 1948 Model C-26 tricycle row-crop was powered by a Continental four-cylinder engine that gave the tractor three-plow performance. Ed Spiess of Rock Island, Illinois, collected this tractor and many lesser-known classics.

1954 Intercontinental C-26
The military market for tractors apparently sustained the post-war sales of tractors made by recently established tractor makers. This Intercontinental C-26 wide–front tractor was made for the U. S. Navy in 1954. Garland, Texas–based Intercontinental Manufacturing equipped the machine with a Continental engine. Lee Black of Forreston, Illinois, collected this classic tractor.

four-cylinder inline engines. Russell's 1917 Little Boss was rated 15/30. The company was sold in 1927, ending manufacture, but the Russell tractors were serviced through 1942.

THE SILVER KINGS

From an Ohio manufacturer of railroad switch engines came one of the fastest tractors of the 1930s: the Plymouth 10/20. Fate-Root-Heath Company of Plymouth, Ohio, produced the Plymouth 10/20 in 1933 and equipped it with rubber tires so it could travel up to 25 miles per hour on the highway. By 1935, the F-R-H tractor line was renamed Silver King and, not surprisingly, sported a silver paint scheme. Tricycle Silver Kings with forward-mounted single wheels began appearing in 1936. These row-crop-ready machines came with cultivator mounts and added rear-axle clearance. They were designated by numbers referring to their rear-wheel tread width in inches (60, 66, and 72).

In 1940, the Silver King designations increased to three numbers, the first two still indicating rear-tire tread width. The later

1944 Leader A
The yellow tricycle row-crop Leader Model A was a product of the Leader Tractor Manufacturing Company of Chagrin Falls, Ohio, which built tractors from 1937 until 1948. Only a dozen copies of the Model A were built. The simple tricycle row-crop tractor uses a Chrysler 231-cubic-inch engine to produce its 44 horsepower. Leader later produced smaller B, C, and D model one-row machines. Henry Hahn of Perryville, Missouri, found this rare tractor in Pennsylvania and brought it back to Missouri to restore.

1948 Long A
This 1948 Long Model A two-plow row-crop tractor was made by the Long Manufacturing Company of Tarboro, North Carolina, to help fill the shortage of tractors caused by World War II. Several post-war tractors—manufactured by previously unknown makers—competed for sales with the popular Farmall Model H. Long used a four-cylinder Continental engine to power its Model A. Lee Black of Forreston, Illinois, owns this tractor.

1915 Russell 30/60
A transverse-mounted 8x10-inch four-cylinder vertical engine powered this 1915 Russell 30/60 model tractor. Russell & Company of Massillon, Ohio, began making threshing machines and steam engines in 1878 and offered its customers gas tractors starting in 1909. This 1915 tractor used a huge circular radiator with tubes for cooling. The University of California-Davis Antique Mechanics Students Club did the restoration, and the big machine is part of the university's large collection of historic tractors and farm machinery.

1918 Samson Sieve-Grip

General Motors Corporation of Pontiac, Michigan, tried to challenge Henry Ford's dominance in the farm tractor business in 1918 with the Samson line of farm tractors. GM bought the Samson Sieve-Grip tractor of Stockton, California, developed the Fordson look-alike Samson Model M, and bought the light Jim Dandy motor cultivator to compete with Ford. GM updated the Sieve-Grip with the installation of a 4.75x6-inch GM four-cylinder truck engine of 12/25 power capabilities. GM dropped the design in 1919. Super tractor collector-restorer Fred Heidrick of Woodland, California, found and restored this historic machine.

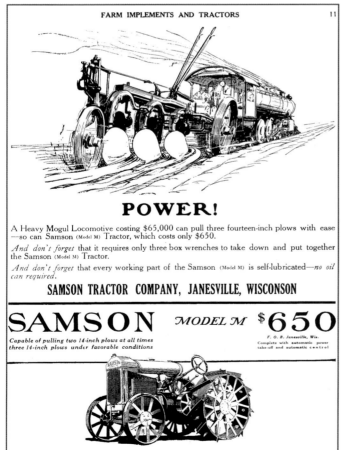

Samson Model M advertisement

tractors also ran faster by 5 miles per hour. By 1956, manufacture of the Silver King had moved to West Virginia, where the tractors were made by Mountain State Fabricating Company of Clarksburg. Production stopped in the late 1950s.

WISCONSIN FARM TRACTORS

Earl McFarland and John Westmount started their tractor company in Sauk City, Wisconsin, in 1917. They soon renamed it the Wisconsin Farm Tractor Company. After making about 600 tractors in three horsepower sizes, the company ceased production in 1923. The Wisconsin tractors were rated at 16/32, 20/35, and 22/40, depending on the engines supplied with the tractors.

1919 Samson M

This 1919 Samson Model M tractor was the machine GM hoped would compete with the very popular Fordson. The Model M's power came from its four-cylinder Northway 276-cubic-inch engine. Its $650 price was competitive with the Fordson's lower price, since the Samson M was at least equal in performance and came with optional equipment that was extra on the Fordson. Randall Knipmeyer of Higginsville, Missouri, found this Samson M.

1950 Global

Built by the Huber Manufacturing Company of Marion, Ohio, after it had quit making its own tractors, this 1950 Global tractor was shipped overseas as part of the Marshall Plan to aid Europe after World War II. This orphan was powered with a four-cylinder Continental 3.875x 5.5-inch diesel engine. Only six to eight Global tractors were built before the line was discontinued. Earl D. Scott of Marysville, Ohio, located and restored this rare tractor.

1949 Sheppard SD-2

The Sheppard diesel tractors of the early 1950s piqued farmer interest in diesel power. The R. H. Sheppard Company of Hanover, Pennsylvania, initially made three sizes of its rubber-tired diesels. Its SD-1 was a one-cylinder 28-cubic-inch engine making 4 horsepower to pull a 12-inch plow. The SD-2 was powered with a 126-cubic-inch two-cylinder 4.25x5-inch engine that made 24 belt horsepower for two-plow power. Its Model SD-3 used three 4x5-inch cylinders and pulled three plows. Its largest tractor, the SD-4 of 1954, had four 4.5x5-inch cylinders and was a four-plow tractor. Sheppard's diesels were available as standard, row-crop, orchard, and industrial models from 1949 to 1956. The company continues in business today as a supplier of truck power steering systems. Jim Bowersox of Lititz, Pennsylvania, restored this 1949 Sheppard SD-2 two-bottom tractor.

1949 Sheppard SD-3

Retired farmer and tractor collector Bob Young of Cambridge, Illinois, restored this 1949 Sheppard Model SD-3 three-plow diesel for his collection. Its three 4x5-inch cylinders create 188 cubic inches of displacement and produce 32 belt horsepower. The Sheppard diesel was designed to be economical and easy to service and maintain. An auxiliary overdrive gave Young's SD-3 a road speed of about 15 miles per hour.

SYMBOLS OF THE PAST

By no means does the above listing cover the wide spectrum of tractor manufacturers active in the 1900 to 1960 period. It is but a brief look at some of the better-known makers of tractors that converted farming from muscle power to motor power. Though memory of the old machines fades with time, their contributions to that revolutionary change in farm power are being preserved. The efforts of thousands of collectors who buy them as rusting hulks, then tenderly and carefully restore them and display them, lets the rest of us appreciate the contribution these old machines made to the bountiful life we all enjoy.

1936 Silver King R-66
First named the Plymouth tractor since it was built in Plymouth, Ohio, this machine was renamed the Silver King after Dodge Motor Company objected to the use of the Plymouth name. Silver King tractors, like this single front wheel Model R-66, were built by the Fate-Root-Heath Manufacturing Company as row-crop tractors and standard-tread machines. Paul Brecheisen of Helena, Ohio, restored this 1936 row-crop Silver King.

Four-Wheel-Drive Power Arrives on the Scene

The largest and most powerful farm tractor built in the twentieth century was the 1977 Big Bud 16V-747. It was designed and built by Northern Manufacturing Company of Havre, Montana, to produce an amazing 760 horsepower. It is currently turning out more than 900 horsepower.

The 16-cylinder turbocharged four-wheel-drive Big Bud tractor was built to the specifications of the Rossi Farms of California. Rossi wanted a machine with big power to break up deep hardpan in the soil so its cotton production would increase. Rossi already was a user of Northern Manufacturing's other models of the Big Bud four-wheel-drive tractors and trusted the company's reputation for making reliable big-horsepower machines.

After the Big Bud 16V-747 arrived on the farm, the Rossi brothers worked the powerhouse hard for eleven years before selling it to a Florida firm. The Big Bud was named 16V-747 for its V-configured 16-cylinder 1,472-cubic-inch two-cycle Detroit Diesel engine and for the Boeing 747 jumbo transport jet of its era—then the largest plane flying. Like its namesake, it was a jumbo tractor.

The thirty-year-old classic is still at work, now for its third owners. Brothers Robert and Randy Williams of Big Sandy, Montana, use it to till 8,000 acres of spring wheat ground on their operation in Choteau County, just south of Montana's Bear Paw Mountain. They bought the Big Bud in 1997 from its second owner, Willowbrook Farms of Indialantic, Florida, where it had

Big Bud 16V-747
The Big Bud 16V-747 was the largest farm tractor made in the twentieth century. It was powered by a 1,472-cubic-inch V-configured 16-cylinder Detroit Diesel engine that now produces more than 900 horsepower. The Big Bud can till about an acre of wheat ground per minute on the Big Sandy, Montana, wheat ranch where it works for owners Robert and Randy Williams. The big diesel engine is turbocharged and aftercooled. Engine power is transmitted through a twin-disc, clutch-free, power-shift transmission. The machine's final drive includes limited-slip differentials with planetary drive. The tractor was built in Havre, Montana, in 1977 and 1978 by Ron Harmon and the crew of the Northern Manufacturing Company.

worked for many years and been retired. The Williams brothers had the behemoth shipped back across the United States to Havre, Montana, where it was built twenty years earlier, for its restoration. The Big Bud now works within sixty miles of where it was built and then rebuilt.

At work in the field, the Big Bud pulls its 80-foot-wide field cultivator at up to 8 miles per hour, working ground at the rate of about one acre per minute. That translates to about sixty acres tilled per hour.

The tractor's physical dimensions are awe-inspiring. The biggest Big Bud stands 14 feet tall to the top of its cab, is 28.5 feet long, and is nearly 21 feet wide, including the width of its dual tires.

The Big Bud four-wheel-drive machines, along with those of other makers, played an important role in the further mechanization of North America, especially in the 1970s in the West.

In 1950 there were no rubber-tired four-wheel-drive tractors tested at Nebraska. They weren't being built. Only the Willys Farm Jeep, the Dodge Power Wagon, and the Range Rover put in an appearance there for tests of four-wheel-drive machines. They didn't work out too well for farm tillage work. Meanwhile, Lloyd Erickson of Elwood, Illinois, known as a mechanical genius, had some ideas for a four-wheel-drive rubber-tired tractor. He tried to interest at least one major tractor maker in manufacturing such a machine. Finding no takers, Erickson's Elwood Engineering Company developed, and, in 1953, made his Big Four 154-horsepower four-wheel-drive diesel tractor. The Big Four was powered with a 572-cubic-inch six-cylinder Continental diesel engine. It had all-wheel steering controlled by a single lever, air brakes, and a five-speed transmission boasting a road speed of up to 15 miles per hour. It turned farmers' heads when they saw it at work. It was a nimble, heavy-pulling farm tractor like none of the others made to that date.

Erickson made eight of the big tractors and sold them all. But he didn't find an interested manufacturer. While his belief that a high-horsepower rubber-tired tractor could give large tractors

Big Bud 16V-747
The 16V-747 rides high on eight 35x38-inch custom-built tires that are eight feet in diameter. The tractor's width is just about 21 feet with the duals. The 16V-747 measures 14 feet tall from the ground to top of the cab. Front and rear axles are 16.25 feet apart, and the tractor stretches out to 28.5 feet in length. Its shipping weight is about 47.5 tons. Ballasted for work, it can weigh up to 65 tons. This Big Bud's fuel load is 1,000 gallons of diesel.

transport mobility that crawler tractors lacked was right, he didn't get to capitalize on his ideas. Over the next twenty-five years, developments in North American farming proved there was a need for four-wheel-drive tractors in the vast open wheat fields of the western states and Canada.

As the four-wheel-drive tractor concept grew in favor, so did the horsepower increases on these machines. Who then could even imagine 700-, 800-, or 900-horsepower tractors?

FOUR-WHEEL-DRIVES FROM PORTLAND AND MINNESOTA

The Wagner brothers of Portland, Oregon, were among the early leaders in the four-wheel-drive field. Their first four-wheel-drive tractor was announced in 1953. Three years later, the Wagner Tractor, Inc. firm offered 57-, 73-, and 96-horsepower four-wheel-drive machines. California observers soon saw the tractors as already eclipsing crawler tractors. Other 1950s-era four-wheel-drive machines include the Harris Power Horse of 1954, a 33-drawbar horsepower skid-steer tractor. It stayed in production until the mid-1960s.

In 1957 and 1958, farmer brothers Douglass and Maurice Steiger built the kind of four-wheel-drive tractor they needed on their northwest Minnesota farm. Made from available components, the tractor ran its 238-horsepower Detroit Diesel engine for some 10,000 hours. The Steigers' neighbors noticed and soon wanted a tractor like the one the brothers had made.

The Steiger's second generation of tractors, Models 1200, 1700, 2200, and 3300, were all built during the 1960s on the Steiger farm. One hundred and twenty were built and sold in the Dakotas, Montana, and in Saskatchewan and Alberta, Canada. The Steiger 3300 used a 318-horsepower Detroit Diesel engine. These weren't toy tractors. They were serious workhorses filling a need for big farm power.

In 1969, Steiger built its new 175-horsepower four-wheel-drive Wildcat at its large new plant in Fargo, North Dakota. The Wildcat was powered by a Caterpillar 3145 V-8 engine. Steiger's 800 Series Tiger also came out that year. Its power source was a Cummins V-903 engine that could produce up to 320 horsepower.

FOUR-WHEEL POWER FROM THE MAJORS

John Deere was the first major U.S. tractor maker to offer a four-wheel-drive tractor. Deere came out with the big 8010 in 1959. It was a large four-wheeler powered with a GM six-cylinder 215-horsepower diesel. The 8010 and an improved 8020 version were offered only until 1964.

International Harvester announced its Model 4300 four-wheel-drive 214.23-drawbar horsepower tractor in 1961. It was the first IH farm tractor to be turbocharged. IH soon realized it had overshot the needed power level and toned down its next similar tractor.

The second IH four-wheel-drive tractor, the Model 4100, came out in 1965. It was a less powerful version of the 4300 shown in 1961. The 4100 used a 429-cubic-inch six-cylinder turbocharged IH engine to make its 110.82-horsepower drawbar pull. Realizing that the four-wheel-drive machine would be used for long hours

of tillage work in cold conditions in early spring and late fall, IH offered it with an enclosed heated and air-conditioned cab. It was the first such cab International factory-installed on its tractors.

In 1964, J. I. Case marketed its first four-wheel-drive, the Model 1200 Traction King Diesel. The 451-cubic-inch six-cylinder turbocharged diesel produced 106.86 drawbar horsepower in Nebraska tests. The Case Model 1200 was that company's first turbocharged farm tractor. Case's second-generation four-wheel-drive tractor debuted in 1969. The Case 1470 Traction King was powered by a larger, 504-cubic-inch, Case six-cylinder turbocharged diesel that produced 132.06 drawbar horsepower.

Minneapolis-Moline soon joined in on the four-wheel-drive trend. Its Model A4T-1600 diesel was offered beginning in 1969. The A4T-1600's six-cylinder 585-cubic-inch engine was good for 127.76 drawbar horsepower at its Nebraska tests. The same machine was later offered as the Oliver Model 2655 diesel. Versatile Manufacturing of Winnipeg, Manitoba, first entered the

Big Bud 16V-747
This Big Bud is well traveled. Built in Montana in the late 1970s, Big Bud 16V-747 worked in southern California's cotton fields for eleven years, then in Florida for about ten years, and retired before 1997. It was then shipped back to Montana for restoration and another go at the working life. A shed for this tractor would be as large as a two-level four-bedroom house.

four-wheel-drive market in 1966 with its Model D-100 diesel. It had a 363-cubic-inch six-cylinder Ford engine with a 125-horsepower rating. By 1967, the Versatile D-118 with 140 horsepower from a 352-cubic inch V6 Cummins diesel had arrived. Four-wheel-drive was catching on.

BIG BUD ARRIVES

The first of the Big Bud tractors from Havre, Montana, was introduced in 1968. The Model HN 250 diesel had an estimated horsepower rating of 310 from an 855-cubic-inch six-cylinder Cummins diesel. The first Big Bud weighed 34,000 pounds and had twelve forward gears.

The replacement for the Big Bud Model 250 was introduced in 1970: the Big Bud Model HM 320 diesel. It had an 855-cubic-inch six-cylinder turbocharged Cummins diesel with intercooler. Its advertised horsepower was 320 and it featured a partial-range power shift transmission.

1953 Big 4 by Erickson
The popular big row-crop tractor when the Big 4 was built was the new Farmall Super MTA with 46 horsepower. The Big 4 could pull six to eight 16-inch bottom plows, or more than twice as many as the Farmall. Big 4 drivers had a comfortable seat and a good view of their surroundings. Erickson made the Big 4 to compete with D-4 and D-6 Caterpillar tracklayer tractors.

MORE POWER FROM THE MAJORS

By 1971, the new Case Model 2470 diesel four-wheel-drive tractor was producing 154.24 drawbar horsepower from its 504-cubic-inch six-cylinder Case turbocharged engine. Steering was changed so that all four wheels could be steered at the same time. John Deere offered its new turbocharged 7020, a four-wheel-drive tractor with an intercooler, that same year. The 7020 coaxed 131.50 drawbar horsepower from its 404-cubic-inch six-cylinder engine and came equipped with an enclosed cab for operator comfort. It could be configured to work as a row-crop.

Two new Massey-Ferguson four-wheel-drive machines were introduced in 1971. The M-F 1500 used a V-8 Caterpillar 573-cubic-inch diesel engine. It put out 152.77 drawbar horsepower. The larger Model 1800 had the larger 636-cubic-inch Caterpillar V-8 and produced 178.7 drawbar horsepower in Nebraska tests. The Model 1500 was the first M-F four-wheel-drive tractor since the General Purpose tractor of 1936.

Allis-Chalmers first four-wheel-drive tractor was its Model 440 articulated-steer diesel of 1972. It was made for Allis by Steiger of Fargo, North Dakota. The Model 440 employed a 555-cubic-inch V-8 Cummins engine. Horsepower output was estimated at 165/208, since it was not tested at Nebraska.

The trend in the later 1970s was for more power, simpler construction, and more enclosed cabs on four-wheel-drive tractors. In Big Sky Country, Big Bud continued adding power as it introduced new tractor models. Its HN 350 diesel put out an astounding 350 horsepower. In 1975, the Big Bud Model KT 450 offered 450 horsepower from a big, 1,150-cubic-inch, six-cylinder turbocharged Cummins diesel. The 500 horsepower level was passed in 1977 with the introduction of the Model KT 525. Its six-cylinder 1,150-cubic-inch turbocharged and intercooled engine produced 525 horsepower.

Then along came the monster 16V-747 Big Bud, with horsepower nearing 800, and tractor history was made.

Today, four-wheel-drive tractors continue to handle an increasing share of the continent's heavy tillage work. Will the new machines arriving on the scene surpass the 1,000-horsepower level? Only time will tell.

1940 Silver King 42
This 1940 Silver King 42 still used a single front wheel in its row-crop design, but had new rounded grille work and dished fenders. It is powered by a four-cylinder Continental 162-cubic-inch 3.25x4-inch engine rated at 20-horsepower at 1,800 rpm. It has a six-volt electrical system powering its lights and electric starter. Perry Jennings of Decatur, Tennessee, restored this tractor that needed a lot of work to add to his sizeable collection of more than thirty-five tractors.

1918 Square Turn

The 1918 Square Turn tractor could stop and turn around in its own tracks. Its three-wheel design let it operate in either direction. An engine-powered lift raised and lowered its three-bottom plow. The tractor was rated at 18/35 from a 5x6.5-inch bore-and-stroke four-cylinder Climax engine. Invented by farmer Albert Kenney and machinist A. J. Colwell of Norfolk, Nebraska, the design and manufacturing rights changed hands a few times before production finally returned to Norfolk. This Square Turn is owned by the Elkhorn Valley Historical Society. It is displayed in the group's museum. Merle Rix of Norfolk, Nebraska, is the driver.

1910s Turner Simplicity advertisement

1917 Turner Simplicity 12/20

The 1917 Turner Simplicity 12/20 tractor was built in Port Washington, Wisconsin, by the Turner Manufacturing Company. It was powered by a 3.75x5.25-inch Waukesha engine. A larger 14/25 model was also available. The Turner firm fell on hard financial times and was liquidated in 1920, with the Simplicity name going to a new company, Simplicity Manufacturing Company. Simplicity built small garden tractors for Montgomery Ward beginning in 1937 and became a major supplier of garden tractors. Allis-Chalmers bought Simplicity in 1965 and sold it in 1983. Harvey Jongeling and Kenneth Hoogestraat of Chancellor, South Dakota, restored this Turner Simplicity tractor completely.

1950 Montgomery Wards Lowther Custom
Montgomery Wards Company of Chicago sold this advanced three- to four-plow tractor made by the Harry A. Lowther Company of Shelbyville, Indiana. This 1950 Custom model was powered by a 56-horsepower six-cylinder Chrysler industrial engine with a fluid drive connecting it to the five-speed transmission. By 1953, the Lowther Custom tractors were being built by Custom Tractor Manufacturing Company of Hustisford, Wisconsin. Russel Miner of Pleasant Plains, Illinois, restored and painted his Wards in 1988. It still has its new look.

1918 Wisconsin 22/40
Only 600 of this 1918 Wisconsin 22/40 standard-tread tractor were built by the Wisconsin Farm Tractor Company of Sauk City, Wisconsin. It was powered by a four-cylinder Beaver engine and weighed about 6,000 pounds. Wisconsin made several versions of this tractor using different engines between 1917 and 1923. Frank Wurth of Freeburg, Illinois, bought this tractor from its original owner in 1977 and restored it.

Bibliography

Some of the books, periodicals, and articles used in researching this book include:

Baumheckel, Ralph, and Kent Borghoff. *International Farm Equipment*. St. Joseph, MI: American Society of Agricultural Engineers, 1997.

Broehl, Wayne G. Jr. *John Deere's Company: A History of Deere & Company and Its Times*. New York: Doubleday, 1984.

The Caterpillar Story. Peoria, IL: Caterpillar, Inc. 1990.

Deere & Company marketing writers. *John Deere Tractors: 1918–1994*. St. Joseph, MI: American Society of Agricultural Engineers, 1994.

Erb, David, and Eldon Brumbaugh. *Full Steam Ahead: J. I. Case Tractors & Equipment, 1842–1955*. St. Joseph, MI: American Society of Agricultural Engineers, 1993.

Gay, Larry. *Farm Tractors: 1975–1995*. St. Joseph, MI: American Society of Agricultural Engineers, 1995.

Gray, R. B. *The Agricultural Tractor: 1855–1950*. St. Joseph, MI: American Society of Agricultural Engineers, 1975.

Johnson, Paul C. *Farm Animals in the Making of America*. De Moines, IA: Wallace-Homestead, 1975.

———. *Farm Inventions in the Making of America*. De Moines, IA: Wallace-Homestead, 1976.

———. *Farm Power in the Making of America*. Des Moines, IA: Wallace-Homestead, 1978.

Larsen, Lester. *Farm Tractors: 1950–1975*. St. Joseph, MI: American Society of Agricultural Engineers, 1981.

Leffingwell, Randy. *The American Farm Tractor*. Osceola, WI: Motorbooks International, 1991.

Letourneau, Peter. *Vintage Case Tractors*. Stillwater, MN: Voyageur Press, 1997.

Macmillan, Don. *The John Deere Tractor Legacy*. Stillwater, MN: Voyageur Press, 2003.

———. *The Big Book of John Deere Tractors*. Stillwater, MN: Voyageur Press, 1999.

Macmillan, Don, and Roy Harrington. *John Deere Tractors & Equipment, 1960–1990*. St. Joseph, MI: American Society of Agricultural Engineers, 1991.

Macmillan, Don, and Russell Jones, *John Deere Tractors & Equipment: 1837–1959*. St. Joseph, MI: American Society of Agricultural Engineers, 1988.

Peterson, Chester Jr., and Rod Beemer. *John Deere New Generation Tractors*. Osceola, WI: MBI Publishing Company, 1998.

Prairie Farmer editors. "Farm Power From Muscle to Motor, Revolution in Rubber, Plows that Made the Prairies, The Better the Fuel . . ." *Prairie Farmer* magazine centennial issue, Jan. 11, 1941.

Pripps, Robert N. *The Big Book of Caterpillar*. Stillwater, MN: Voyageur Press, 2000.

———. *Vintage Ford Tractors*. Stillwater, MN: Voyageur Press, 1997.

Swinford, Norm. *Allis-Chalmers Farm Equipment: 1914–1985*. St. Joseph, MI: American Society of Agricultural Engineers, 1994.

Wendel, C. H. *Encyclopedia of American Farm Tractors*. Osceola, WI: Motorbooks International, 1992.

———. *International Harvester: 150 Years*. Osceola, WI: Motorbooks International, 1993.

———. *Minneapolis-Moline Tractors: 1870–1969*. Osceola, WI: Motorbooks International, 1990.

Wik, Reynold M. *Benjamin Holt & Caterpillar: Tracks & Combines*. St. Joseph, MI: American Society of Agricultural Engineers, 1984.

ABOUT THE AUTHOR

Photographer and author Ralph W. Sanders has been taking photographs of tractors for more than sixty years. His first photos were snapshots he took of the tractors he drove as a boy on the grain farm where he was born and raised near Stonington, Illinois. Some of those tractors included a Farmall F-12, Caterpillar R2, Ford-Ferguson 2N, Farmall C, and Caterpillar D2. The harvest of winter wheat in the summer and soybeans in the fall brought out a Stockton, California–built 1929 Holt combine, on which Ralph and his brothers John Jr. ("Jack") and Jim took turns as "header tender." Summers also gave the Sanders boys a chance to bale hay on their grandfather Wetzel's Case manual wire-tie pickup baler.

Rural electrification had just added lights to the Old Stonington one-room country schoolhouse when Ralph started first grade there in 1939. By the time he completed the eighth grade, World War II was over and the Atomic Age had dawned. It was during this time he was captivated by photography as he watched developing prints magically appear in trays spread out on the table in the darkened farmhouse's kitchen.

By the time Ralph graduated from Stonington Community High School in 1951, the Korean War was underway. After studying agriculture at the University of Illinois for three years, Ralph was drafted into the U.S. Army and spent most of his two years of service as a clerk in an ordnance recovery and reclassification company in Germany. With the help of the G.I. Bill, he completed a bachelor's degree in journalism at the University of Illinois and earned his "wings"—his private pilot's license.

His journalism career started with news writing and broadcasting duties at radio stations WCRA in Effingham, Illinois, and then WDZ in Decatur, Illinois. That experience led to daily newspaper reporting with the Decatur, Illinois, *Herald & Review*, farm reporting as Illinois field editor for *Prairie Farmer* magazine, and in 1968, more farm coverage as an associate editor at *Successful Farming* magazine in Des Moines, Iowa.

In 1974, Ralph pursued his passion for photography and became a full-time freelance photographer. He then dusted off his private pilot's license and by 1978 was flying Sanders Photographics' Piper Cherokee Archer II airplane over the eastern half of the United States on assignments for a growing number of

Copyright 2006, Rich Sanders

agricultural accounts. The farm crisis of the 1980s effectively shut down "Sanders Air" and, in 1986, Ralph sold the airplane and reluctantly returned to the highways for travel.

Over the years, Ralph's agricultural photographic work has included assignments for *Successful Farming*, Deere & Company, Massey-Ferguson, Kinze Manufacturing, Vermeer Manufacturing, DuPont Ag, Monsanto Ag, Voyageur Press, and many others.

Ralph photographed the first DuPont *Classic Farm Tractors* calendar in 1989. He continued photographing that calendar for more than eleven years. Through 2005, Ralph photographed two different farm tractor calendars each year for Voyageur Press. The lovingly restored tractors, originally photographed for these calendars, form the basis of the photo illustrations in this book. Conversations with the tractor owners during the creation of the photographs put Ralph on track to preserve some of our country's precious rural heritage—that of the American farm tractor.

Ralph and Joanne, his wife of nearly fifty years, have seven grown children and fifteen grandchildren. They live in West Des Moines, Iowa.

This is Ralph's fourth book with Voyageur Press. The previous books are *Vintage Farm Tractors, Vintage International Harvester Tractors*, and *Ultimate John Deere*.

Index